Springer Proceedings in Physics

Volume 216

The series Springer Proceedings in Physics, founded in 1984, is devoted to timely reports of state-of-the-art developments in physics and related sciences. Typically based on material presented at conferences, workshops and similar scientific meetings, volumes published in this series will constitute a comprehensive up-to-date source of reference on a field or subfield of relevance in contemporary physics. Proposals must include the following:

- name, place and date of the scientific meeting
- a link to the committees (local organization, international advisors etc.)
- scientific description of the meeting
- list of invited/plenary speakers
- an estimate of the planned proceedings book parameters (number of pages/ articles, requested number of bulk copies, submission deadline).

More information about this series at http://www.springer.com/series/361

Yafang Han
Editor

Physics and Techniques of Ceramic and Polymeric Materials

Proceedings of Chinese Materials Conference 2018

Springer

Editor
Yafang Han
Chinese Materials Research Society
Beijing, China

ISSN 0930-8989 ISSN 1867-4941 (electronic)
Springer Proceedings in Physics
ISBN 978-981-13-5946-0 ISBN 978-981-13-5947-7 (eBook)
https://doi.org/10.1007/978-981-13-5947-7

Library of Congress Control Number: 2018966387

This Springer imprint is published by the registered company Springer Nature Singapore Pte Ltd.
The registered company address is: 152 Beach Road, #21-01/04 Gateway East, Singapore 189721, Singapore

Preface

This is the proceedings of the selected papers presented at Chinese Materials Conference 2018 (CMC 2018) held in Xiamen City, Fujian, China, July 12–16, 2018.

The Chinese Materials Conference (CMC) is the most important serial conference of Chinese Materials Research Society (C-MRS) and is held each year since the early 1990s. Chinese Materials Conference 2018 had 35 Symposia covering four fields of energy and environmental materials, advanced functional materials, high-performance structural materials, and design, preparation, and characterization of materials. More than 5500 participants attended the conference, and the organizers received more than 500 technical papers. By recommendation of symposium organizers and after peer reviewing 94 papers are published in the present proceedings, which are divided into two volumes of

Part 1: Physics and Engineering of Metallic Materials and
Part 2: Physics and Techniques of Ceramic and Polymeric Materials

This is the volume for Part 2 including 24 papers selected from 3 symposia of advanced ceramics materials, modification and composite technology for fiber materials, and polymer materials.

The editor would like to give the thanks to the symposium chairs and all the paper reviewers of this volume, especially to Prof. Yafang Han, Prof. Xinqing Zhao, Prof. Jiangbo Sha, Prof. Chungen Zhou, Prof. Wenlong Xiao, and Prof. Tong Liu for their English polishing of the manuscripts.

Beijing, China Yafang Han

Contents

Contributors

Aiming Bu Department of Materials Science and Engineering, Beijing Institute of Technology, Beijing, China

Zhaodong Cao Shanghai Institute of Laser Plasma, China Academy of Engineering Physics, Shanghai, China

Bin Chen The 2nd Pipeline Construction Limited Company of China National Petroleum Corporation, Xuzhou, Jiangsu, China

Weiwei Chen Department of Materials Science and Engineering, Beijing Institute of Technology, Beijing, China

Wenhao Chen Xi'an University of Technology, Xi'an, China

Huanwu Cheng Department of Materials Science and Engineering, Beijing Institute of Technology, Beijing, China

Miao Cheng State Key Laboratory for Modification of Chemical Fibers and Polymer Materials, College of Material Science and Engineering, Donghua University, Shanghai, China

Renju Cheng Chongqing Academy of Science and Technology, Chongqing, People's Republic of China

Shen Cong CNPC Tubular Goods Research Institute, Xi'an, Shaanxi, China

Hanwu Dong Chongqing Academy of Science and Technology, Chongqing, People's Republic of China

Hui Gao College of Materials Science and Engineering, Shandong University of Science and Technology, Qingdao, China

Feng He State Key Laboratory of Silicate Materials for Architectures, School of Materials Science and Engineering, Wuhan University of Technology, Wuhan, China

Xiyun He Shanghai Institute of Ceramics, Chinese Academy of Sciences, Shanghai, China

Shuo Hu State Key Laboratory for Modification of Chemical Fibers and Polymer Materials, College of Material Science and Engineering, Donghua University, Shanghai, China

Zexu Hu State Key Laboratory for Modification of Chemical Fibers and Polymer Materials, College of Materials Science and Engineering, Donghua University, Shanghai, China

Xiaowen Huang College of Materials Science and Engineering, Shandong University of Science and Technology, Qingdao, China

Xin-Lei Jia China North Nuclear Fuel Co. Ltd., Baotou, China

Bin Jiang College of Materials Science and Engineering, Chongqing University, Chongqing, People's Republic of China

Xin-Hao Jiang China North Nuclear Fuel Co. Ltd., Baotou, China

Shi-Dong Kang China North Nuclear Fuel Co. Ltd., Baotou, China

Linggen Kong Jiangsu Haiyang Chemical Fiber Co., Ltd., Taizhou, China

Congcong Li Tianjin Key Laboratory of Advanced Fibers and Energy Storage, Tianjin Polytechnic University, Tianjin, People's Republic of China

Haili Li Chongqing Academy of Science and Technology, Chongqing, People's Republic of China;
College of International Business, Sichuan International Studies University, Chongqing, People's Republic of China

Huairui Li International Centre for Bamboo and Rattan, Beijing, People's Republic of China

Na Li CNPC Tubular Goods Research Institute, Xi'an, Shaanxi, China

Peixian Li School of Materials and Energy, Guangdong University of Technology, Guangdong, Guangzhou, China

Qian Li State Key Laboratory of Advanced Special Steels, School of Materials Science and Engineering, Shanghai University, Shanghai, China;
School of Materials and Energy, Guangdong University of Technology, Guangdong, Guangzhou, China

Yunkai Li Beijing Institute of Technology, Beijing, China

Z. M. Li State Key Laboratory for Modification of Chemical Fibers and Polymer Materials, College of Materials Science and Engineering, Donghua University, Shanghai, People's Republic of China;
AccuPath Medical (Jiaxing) Co. Ltd., Shanghai, China

Liang Ling Shanghai Institute of Ceramics, Chinese Academy of Sciences, Shanghai, China

Haitao Liu Science and Technology on Advanced Ceramic Fibers and Composites Laboratory, College of Aerospace Science and Engineering, National University of Defense Technology, Changsha, China

Jiaming Liu State Key Laboratory for Modification of Chemical Fibers and Polymer Materials, College of Material Science and Engineering, Donghua University, Shanghai, China

Jian-Cheng Liu China North Nuclear Fuel Co. Ltd., Baotou, China

Wenjun Liu College of Materials Science and Engineering, Chongqing University, Chongqing, People's Republic of China

Xiaoqing Liu Center of Materials Research and Testing, Wuhan University of Technology, Wuhan, China

Yuewei Liu State Key Lab for Modification of Chemical Fibers and Polymer Material, College of Materials Science and Engineering, Donghua University, Shanghai, China

Zhaowei Liu Xi'an University of Technology, Xi'an, China

Cai-hong Lu CNPC Tubular Goods Research Institute, Xi'an, Shaanxi, China

Chen Lu National Engineering Laboratory for Textile Fiber Materials and Processing Technology, Zhejiang Sci-Tech University, Hangzhou, Zhejiang, China

Jian Lu State Key Lab for Modification of Chemical Fibers and Polymer Material, College of Materials Science and Engineering, Donghua University, Shanghai, China

Zhenghua Lu School of Materials and Energy, Guangdong University of Technology, Guangdong, Guangzhou, China

Chi Ma Jiangsu Haiyang Chemical Fiber Co., Ltd., Taizhou, China

Mengchen Mao Xi'an University of Technology, Xi'an, China

Xuefeng Mao National Engineering Laboratory for Textile Fiber Materials and Processing Technology, Zhejiang Sci-Tech University, Hangzhou, Zhejiang, China

Ying Meng China North Nuclear Fuel Co. Ltd., Baotou, China

Mingda Ni State Key Laboratory for Modification of Chemical Fibers and Polymer Materials, College of Materials Science and Engineering, Donghua University, Shanghai, China

Weinan Pan State Key Laboratory for Modification of Chemical Fibers and Polymer Materials, College of Materials Science and Engineering, Donghua University, Shanghai, China

Daochun Qin International Centre for Bamboo and Rattan, Beijing, People's Republic of China

M. L. Qin State Key Laboratory for Modification of Chemical Fibers and Polymer Materials, College of Materials Science and Engineering, Donghua University, Shanghai, People's Republic of China;
AccuPath Medical (Jiaxing) Co. Ltd., Shanghai, China

Zongyi Qin State Key Laboratory for Modification of Chemical Fibers and Polymer Materials, Donghua University, Shanghai, China

Pingsun Qiu Shanghai Institute of Ceramics, Chinese Academy of Sciences, Shanghai, China

Ting-ting Qu CNPC Tubular Goods Research Institute, Xi'an, Shaanxi, China

Shuxia Ren Hebei Provincial Key Laboratory of Traffic Engineering Materials, School of Materials Science and Engineering, Shijiazhuang Tiedao University, Shijiazhuang, China

Zhongming Ren State Key Laboratory of Advanced Special Steels, School of Materials Science and Engineering, Shanghai University, Shanghai, China

Jiaguang Sheng Jiangsu Haiyang Chemical Fiber Co., Ltd., Taizhou, China

Jiang Shi State Key Laboratory of Silicate Materials for Architectures, School of Materials Science and Engineering, Wuhan University of Technology, Wuhan, China

Liang Song College of Materials Science and Engineering, Shandong University of Science and Technology, Qingdao, China

Peng Song Pipeline Research Institute of CNPC, Langfang, Hebei, China

Bin Sun State Key Lab for Modification of Chemical Fibers and Polymer Material, College of Materials Science and Engineering, Donghua University, Shanghai, China

Yufei Tang Xi'an University of Technology, Xi'an, China

Z. H. Tang School of Materials Science and Engineering, University of Shanghai for Science and Technology, Shanghai, China

Xiujuan Tian College of Materials Science and Engineering, Shandong University of Science and Technology, Qingdao, China

Ke Tong CNPC Tubular Goods Research Institute, Xi'an, Shaanxi, China

Chao Wang PetroChina West East Gas Pipeline Company, Pudong, Shanghai, China

Lu Wang Department of Materials Science and Engineering, Beijing Institute of Technology, Beijing, China

Xiaoqing Wang Tianjin Polytechnic University, Tianjin, People's Republic of China

Xiuhua Wang National Engineering Laboratory for Textile Fiber Materials and Processing Technology, Zhejiang Sci-Tech University, Hangzhou, Zhejiang, China

Zhi Wang Hebei Provincial Key Laboratory of Traffic Engineering Materials, School of Materials Science and Engineering, Shijiazhuang Tiedao University, Shijiazhuang, China

Zhongwei Wang College of Materials Science and Engineering, Shandong University of Science and Technology, Qingdao, China

Ali Wen China Institute of Atomic Energy, Beijing, China

Canqing Wu National Engineering Laboratory for Textile Fiber Materials and Processing Technology, Zhejiang Sci-Tech University, Hangzhou, Zhejiang, China

Chao Wu Beijing Institute of Technology, Beijing, China

Qiguang Wu Analysis and Test Center, Guangdong University of Technology, Guangzhou, China

Qilin Wu State Key Lab for Modification of Chemical Fibers and Polymer Material, College of Materials Science and Engineering, Donghua University, Shanghai, China

Peng Xi Tianjin Key Laboratory of Advanced Fibers and Energy Storage, State Key Laboratory of Separation Membranes and Membrane Processes, Tianjin Polytechnic University, Tianjin, People's Republic of China

Hengxue Xiang State Key Laboratory for Modification of Chemical Fibers and Polymer Materials, College of Materials Science and Engineering, Donghua University, Shanghai, China

Xinyu Xiang Beijing Institute of Technology, Beijing, China

Yan Xiang Department of Materials Science and Engineering, Beijing Institute of Technology, Beijing, China

Junlin Xie State Key Laboratory of Silicate Materials for Architectures, School of Materials Science and Engineering, Wuhan University of Technology, Wuhan, China

Weijie Xin College of Materials Science and Engineering, Shandong University of Science and Technology, Qingdao, China

Rong Xu Xi'an University of Technology, Xi'an, China

Y. Y. Xue School of Materials Science and Engineering, University of Shanghai for Science and Technology, Shanghai, China

Xuhuan Yan Tianjin Polytechnic University, Tianjin, People's Republic of China

Hu Yang School of Materials Science and Engineering, Wuhan University of Technology, Wuhan, China

Liu Yang Chongqing Academy of Science and Technology, Chongqing, People's Republic of China

Xiaoqing Yang International Centre for Bamboo and Rattan, Beijing, People's Republic of China

Zhigang Yang Hebei Provincial Key Laboratory of Traffic Engineering Materials, School of Materials Science and Engineering, Shijiazhuang Tiedao University, Shijiazhuang, China;
State Key Laboratory of Advanced Special Steels, School of Materials Science and Engineering, Shanghai University, Shanghai, China

Dengzun Yao Pipeline Research Institute of CNPC, Langfang, Hebei, China

Jianbo Yu State Key Laboratory of Advanced Special Steels, School of Materials Science and Engineering, Shanghai University, Shanghai, China

M. H. Yu State Key Laboratory for Modification of Chemical Fibers and Polymer Materials, College of Materials Science and Engineering, Donghua University, Shanghai, People's Republic of China

Qing Yu College of Materials Science and Engineering, Shandong University of Science and Technology, Qingdao, China

Daqing Yuan China Institute of Atomic Energy, Beijing, China

Gecheng Yuan School of Materials and Energy, Guangdong University of Technology, Guangdong, Guangzhou, China

Xianfeng Yue International Centre for Bamboo and Rattan, Beijing, People's Republic of China

Fanxin Zeng State Key Laboratory for Modification of Chemical Fibers and Polymer Materials, College of Material Science and Engineering, Donghua University, Shanghai, China

Xia Zeng Shanghai Institute of Ceramics, Chinese Academy of Sciences, Shanghai, China

Gongxun Zhai State Key Laboratory for Modification of Chemical Fibers and Polymer Materials, College of Materials Science and Engineering, Donghua University, Shanghai, China

Na Zhang Chongqing Academy of Science and Technology, Chongqing, People's Republic of China

Rong Zhang International Centre for Bamboo and Rattan, Beijing, People's Republic of China

Xiaoshan Zhang Science and Technology on Advanced Ceramic Fibers and Composites Laboratory, College of Aerospace Science and Engineering, National University of Defense Technology, Changsha, China

Xuzhen Zhang National Engineering Laboratory for Textile Fiber Materials and Processing Technology, Zhejiang Sci-Tech University, Hangzhou, Zhejiang, China

Yanyan Zhang Qingdao Technological University Qindao College, Qingdao, China

Yongfu Zhang Department of Materials Science and Engineering, Beijing Institute of Technology, Beijing, China

Youwei Zhang College of Material Science and Engineering, Donghua University, Shanghai, China

Yuping Zhang Department of Materials Science and Engineering, Beijing Institute of Technology, Beijing, China

Yuxi Zhang College of Materials Science and Engineering, Shandong University of Science and Technology, Qingdao, China

Kang Zhao Xi'an University of Technology, Xi'an, China

Suchuan Zhao The Science Academy, Shanghai University, Shanghai, China

Tianxiang Zhao Tianjin Key Laboratory of Advanced Fibers and Energy Storage, Tianjin Polytechnic University, Tianjin, People's Republic of China

Xinyi Zhao The Fourth Military Medical University, Xi'an, China

Zhijia Zhao Hebei Provincial Key Laboratory of Traffic Engineering Materials, School of Materials Science and Engineering, Shijiazhuang Tiedao University, Shijiazhuang, China

Zhe Zhou State Key Lab for Modification of Chemical Fibers and Polymer Material, College of Materials Science and Engineering, Donghua University, Shanghai, China

Bin Zhu The Science Academy, Shanghai University, Shanghai, China
Shanghai Institute of Ceramics, Chinese Academy of Sciences, Shanghai, China

Jiliang Zhu College of Materials Science and Engineering, Sichuan University, Chengdu, China

Meifang Zhu State Key Laboratory for Modification of Chemical Fibers and Polymer Materials, College of Materials Science and Engineering, Donghua University, Shanghai, China

S. Zhu State Key Laboratory for Modification of Chemical Fibers and Polymer Materials, College of Materials Science and Engineering, Donghua University, Shanghai, People's Republic of China

Preparation of Al_2O_3 Ceramic Cores by Dry-Pressing Assisted of Precursor-Derived Ceramic Technology

Zhigang Yang, Zhijia Zhao, Qian Li, Jianbo Yu, Zhongming Ren, Zhi Wang and Shuxia Ren

Abstract Porous Al_2O_3 ceramic cores were prepared successfully by dry-pressing assisted of precursor-derived ceramic technology. Among, Al_2O_3 powders with flake shape were used as the matrix and silicone resin powders were used as the precursor. The effects of sintering temperatures on the properties of Al_2O_3 ceramic cores were researched. Results showed that the silicone resin powders had well-cohesive property, making the Al_2O_3 grains bond closely under pressure. During the sintering, the pyrolysis of silicone resin happened and the amorphous silica formed firstly at low temperature. The increasing temperatures made the amorphous silica gradually crystallize. And when the temperature was more than 1500 °C, the silica reacted with Al_2O_3 to form the mullite phases. With the increase of temperature, the shrinkage rate reached the maximum value of 0.95%, and the bulk density reached the maximum value of 2.53 g cm^{-3}, and the porosity reached the minimum value of 33.11%, and the bending strength had the highest value of 105 MPa at 1500 °C.

1 Introduction

Ceramic cores with excellent properties such as the mechanical strength, thermal stability, dimensional stability, and thermal shock resistance are widely researched in the casting field to fabricate the hollow parts with precise size [1, 2]. To date, two kinds of ceramic cores have been widely used in the production of hollow parts in the industry, including the silica-based and alumina-based ceramic cores. In silica ceramic cores, the complex phase transformation at different temperatures happens. For example, the formed cristobalite during sintering plays a role in the strength of

Z. Yang · Z. Zhao · Z. Wang · S. Ren
Hebei Provincial Key Laboratory of Traffic Engineering Materials, School of Materials Science and Engineering, Shijiazhuang Tiedao University, Shijiazhuang 050043, China

Z. Yang · Q. Li · J. Yu (✉) · Z. Ren
State Key Laboratory of Advanced Special Steels, School of Materials Science and Engineering, Shanghai University, Shanghai 200072, China
e-mail: jbyu@shu.edu.cn

Y. Han (ed.), *Physics and Techniques of Ceramic and Polymeric Materials*,
Springer Proceedings in Physics 216,
https://doi.org/10.1007/978-981-13-5947-7_1

ceramics. But a large number of cristobalite will lead to the appearance of cracks in the ceramic cores, as a result, decreasing the properties. The effective control of cristobalite content is very difficult in the industry. Additionally, when the used temperature is above than 1550 °C in the casting of metal parts, the thermal stability of silica-based ceramic cores decreases and the creep deformation becomes heavy, leading to their deformation and softening [3, 4]. If so, the dimension and quality of hollow parts will be impacted severely. Therefore, the silica-based ceramic cores are not used in the casting of single crystal and eutectic crystal parts at higher temperature. Differently, compared with silica ceramic cores, Al_2O_3 has higher thermal stability and strength at high temperature. During the sintering, no phase transformation occurs and the phase structure is stability in Al_2O_3 ceramics. Meanwhile, the thermal expansion coefficient matches with that of ceramic shell. Importantly, the Al_2O_3 ceramic cores can be used at the temperature range of 1520–1875 °C [5, 6]. Therefore, the development of Al_2O_3 ceramic cores is essential for the improvement of hollow parts with excellent properties.

Up to date, many methods have been developed to fabricate the ceramic cores. Among them, the precursor-derived ceramic technology has caused the wide attention from researchers owing to its well characteristic such as the processing versatility, lower fabricating temperature and higher mechanical properties [7]. With the increase of temperature, the organic–inorganic transformation of precursor will happen and the ceramic phase will be formed, which is help for the improvement of mechanical properties of the material when the precursor ceramics are mixed with other ceramic matrix. Meanwhile, the precursor will release many gases, which is beneficial for the formation of open-pore structure in samples. In a word, the precursor-derived ceramic technology is very effective for preparing the porous ceramics with excellent properties. For this reason, we plan to use this technology to prepare porous ceramic cores. At first, we have prepared the silica-based ceramic cores and alumina-based ceramic cores with good properties by quartz or alumina powders and solvent silicone resin polymer [8, 9]. Then, we find that the silicone resin powders have good cohesive property, which is beneficial for the shaping of ceramic cores with special shapes by pressing.

In present research, alumina-based ceramic cores were prepared by dry-pressing. Alumina powders were used as the matrix, and silicone resin powders were used as the binder and precursor. The effects of sintering temperatures on the properties of alumina-based ceramic cores were studied.

2 Experimental Procedure

Alumina powders with flake shapes were used as the matrix materials. The average diameter was about 4 μm. The silicone resin powders with Si-O-Si as the main chain were used as the binder and precursor. Firstly, the alumina powders were mixed with silicone resin powders at the weight ratio of 95:5. The mixture was ball-milled for six hours in the plastic bottle using ZrO_2 balls as the milling medium. After the mixture was sieved by 100 mesh, the samples were dry-pressed into the rectangular shape of 4 mm × 5 mm × 45 mm using the metal mold. The holding pressure time was about 3 min. Afterward, the pressed samples were firstly cross-linked at 250 °C for 2 h with heating rate of 2 °C/min, and then sintered at different temperatures for 2 h with 5 °C/min heating rate in air atmosphere. The sintering temperatures were 1300, 1400, 1500, and 1600 °C, respectively.

The linear shrinkage rate of sintered samples was calculated by measuring the dimension before and after sintering. The apparent porosity and bulk density were calculated by the Archimedes' method in distilled water. The water with the samples was heated at 100 °C for 4 h. The phase composition of sintered samples was analyzed by D/max-2500 X-ray diffraction (XRD). The microstructure of sintered samples was examined by SU-1500 scanning electron microscopy (SEM). Then the bending strength of sintered samples at room temperature was measured by three-point bending machine (WDW-300, China). Its span length was about 30 mm. The loading rate was 0.5 mm/min. The final result was the average of five samples.

3 Results and Discussion

Figure 1 shows the XRD patterns of prepared porous Al_2O_3 ceramic cores at different sintering temperatures. It is seen that when the temperature is 1300 and 1400 °C, the samples consist of only Al_2O_3 phase. However, when the temperature is more than 1500 °C, the samples mainly consist of Al_2O_3 phase and a small content of mullite phase. With the increase of temperature, the content of mullite phase gradually increases. It is reported based on previous studies that the silicone resin with Si-O-Si as main chain will firstly form the amorphous silica when the temperature is below than 1300 °C. The further increasing sintering temperature makes the amorphous silica gradually crystallization, leading to the formation of cristobalite [9]. Owing to the amorphous state of silica and low content of cristobalite compared with stronger peaks of Al_2O_3 phase, their weak peak value cannot be observed at 1300 and 1400 °C. At 1500 °C, the formed silica reacts with Al_2O_3 phase as the matrix to form mullite phase, leading to the appearance of mullite phase. The chemical reaction can be promoted by the activation energy. The increasing temperatures offer more activation energy for the reaction between the silica and Al_2O_3. Therefore, more mullite phase forms with the temperature rising. These results are consistent well with that in our early study about the Al_2O_3 ceramic prepared by solvent silicone resin.

Figure 2a shows the shrinkage rate of porous Al_2O_3 ceramic cores via sintering temperature. It is clearly seen that with the increase of sintering temperature, the shrinkage rate of the samples gradually increases. When the temperature is at 1500 °C, the shrinkage rate reaches to the maximal value of 0.95%. When the temperature is more than 1500 °C, the shrinkage rate gradually decreases. Silicone resin with Si-O-Si as main chain will experience the pyrolysis process, producing more gases from the samples [10]. These gases will restrain the shrinkage caused by the sintering driving force. As a result, during the whole sintering, the shrinkage rate of the samples is still

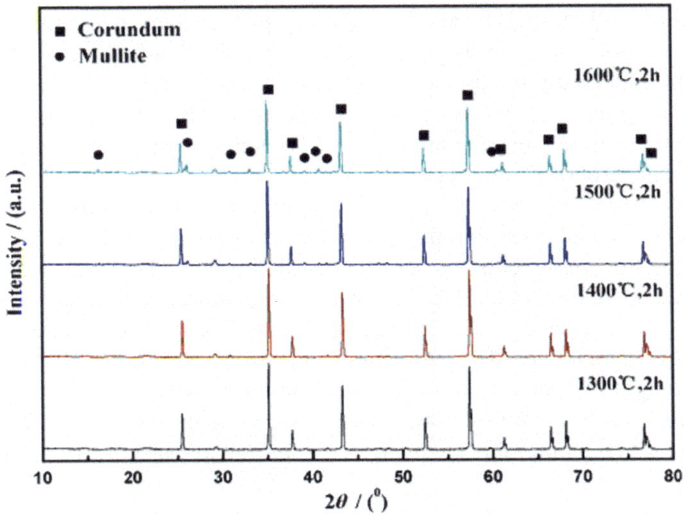

Fig. 1 XRD patterns of porous Al_2O_3 ceramic cores at different temperatures

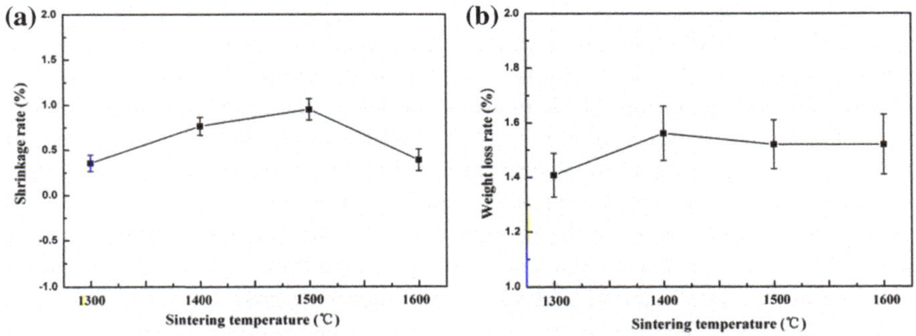

Fig. 2 **a** Shrinkage rate of porous Al_2O_3 ceramic cores in different sintering temperatures, **b** weight loss of porous Al_2O_3 ceramic cores in different sintering temperatures

small. When the temperature increases from 1300 to 1500 °C, the shrinkage is mainly caused by the sintering driving force because the release of gases has completed. At the temperature which is more than 1500 °C, the expansion of the samples happens due to the formation of new phase, which causes the shrinkage decrease.

Figure 2b shows the weight loss of the porous Al_2O_3 ceramic cores via sintering temperatures. It is seen that with the increase of sintering temperature from 1300 to 1400 °C, the weight loss gradually increases from 1.41 to 1.56%. The further increasing temperature can not make the weight loss change evidently. The weight loss of the sample is mainly caused by the pyrolysis of silicone resin. During low temperature below than 1000 °C, the silicone resin with Si-O-Si as the main chain

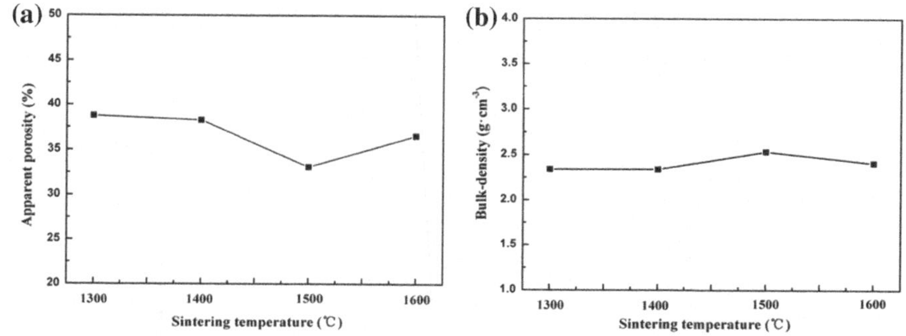

Fig. 3 **a** Apparent porosity, **b** Bulk density of porous Al_2O_3 ceramic cores via sintering temperature

will release many gases owing to the crosslinking and pyrolysis process, leading to the weight loss of the samples. These reasons include the degradation of main chain and decomposition of side chains. When the temperature is above than 1300 °C, the impurity phase and formed liquid phase may volatilize, causing the slightly decreasing of weight.

Figure 3 shows the apparent porosity and bulk density of porous Al_2O_3 ceramic cores via sintering temperatures. It is seen from Fig. 3a that with the increase of the sintering temperature from 1300 to 1500 °C, the apparent porosity gradually decreases from 38.7 to 33.1%. At the temperature of 1500 °C, the apparent porosity reaches the minimum value. On the contrary, with the further increasing temperature, the apparent porosity of the sample increases. In Fig. 3b, it is seen that at the temperature of 1500 °C, the bulk density reaches the maximum value of 2.53 g cm^{-3}. When the temperature is more than or below 1500 °C, the bulk density decreases. These variations are mainly changed along with the shrinkage change via sintering temperatures. The big shrinkage causes the small apparent porosity and big bulk density in the sample.

Figure 4 shows the bending strength of porous Al_2O_3 ceramic cores via sintering temperature. It is found that with the increasing temperature from 1300 to 1500 °C, the bending strength of the samples gradually increases from 92 to 105 MPa. When the sintering temperature is 1500 °C, the bending strength reaches the maximum value. The further increasing of sintering temperature promotes the bending strength decrease. At 1500 °C, the higher bulk density and lower apparent porosity lead to the highest bending strength in the samples. When the sintering temperatures are below or more than 1500 °C, the bigger apparent porosity makes the bending strength evidently decrease.

The SEM fracture pictures of porous Al_2O_3 ceramic cores in different sintering temperature are showed in Fig. 5. In Fig. 5a–d; it is seen that the Al_2O_3 particles are connected by the products formed by the pyrolysis of silicone resin. But many pores with open structure are still clearly seen, showing the formation of the porous ceramic cores. When the Al_2O_3 particles are mixed with silicone resin powders,

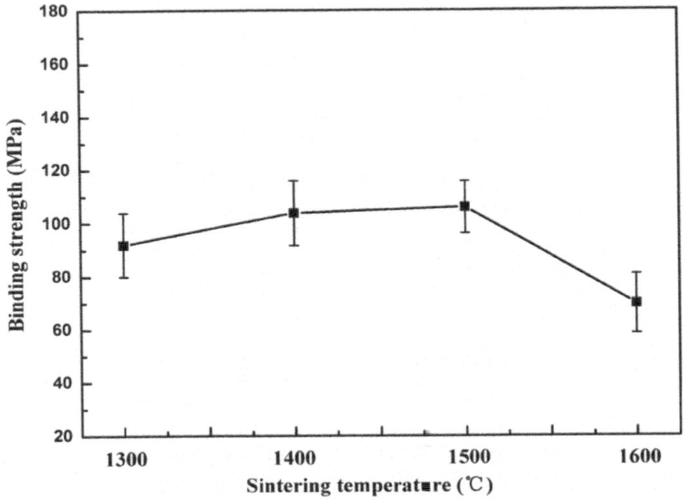

Fig. 4 Bending strength of porous Al_2O_3 ceramic cores ▾ia sintering temperature at room temperature

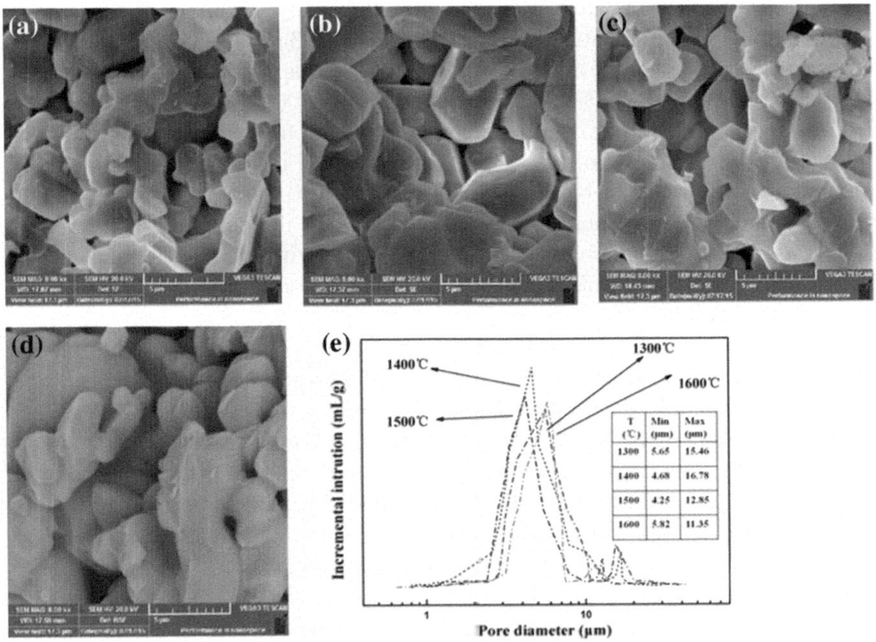

Fig. 5 SEM images of sample sintered at **a** 1300 °C, **b** 1400 °C, and **c** 1500 °C, BSE images, **d** porous Al_2O_3 ceramic cores formed at 1600 °C, and **e** pore size distribution of porous alumina ceramic cores

firstly, the Al_2O_3 particles are bonded by silicone resin powders under pressure. With the increasing temperatures, the crosslinking process of silicone resin powders occurs at temperature of 250 °C, leading to the increase of the connecting strength between the particles. With the further increase of temperatures, the silica by the pyrolysis of silicone resin mainly connects the Al_2O_3 particles. When the temperature is more than 1500 °C, the reaction between the silica and Al_2O_3 particles happens to form the mullite phase, connecting the Al_2O_3 particles. The increasing temperature makes the connection between the Al_2O_3 particles gradually tighter, leading to the improvement of bending strength. At 1600 °C, the decreasing of bending strength is from two aspects: the increasing of porosity and the weakness of connection between the particles caused by the reaction. Figure 5e shows the pore size distribution of porous Al_2O_3 ceramic cores at different temperatures. It is seen that in all the ceramic cores, the pore size shows a bimodal peak distribution, where the size of most of the pores is small and the size of some pores is big. The increasing temperature from 1300 to 1500 °C promotes the decrease of pore size. On the contrary, at 1600 °C, the pore size increases slightly in ceramic cores owing to the decrease of shrinkage.

In the preparation of ceramic cores, the sintering temperature plays a decisive role in the morphology, further influencing the properties of ceramic cores. With the increase of the sintering temperature, the sintering necks between the Al_2O_3 particles are formed by the viscous fluid and the distance between the particles gradually becomes small, leading to the decrease of pore size. As a result, the connecting strength between the Al_2O_3 particles increases, causing the increase of bending strength owing to the fracture energy consumption. At 1600 °C, due to the formation of more mullite phase, the shrinkage of ceramic cores becomes smaller and the pores with big size are formed. The strength of sintering necks between the particles becomes weak, leading to the decrease of the bending strength. Therefore, the proper sintering temperature is very important for the improvement of bending strength of Al_2O_3 ceramic cores.

4 Conclusions

Porous Al_2O_3 ceramic cores were prepared successfully by the precursor-derived ceramic technology. The silicone resin powders with well-cohesive property were used as the precursor and binder. With the increasing of temperature, the silicone resin firstly transformed into amorphous silica at low temperature, and then formed the cristobalite at high temperature. In the Al_2O_3 samples, the silica formed by silicone resin reacted with Al_2O_3 grains to form the new mullite phase. The sintering temperature had an important role in the properties of Al_2O_3 ceramic cores. At 1500 °C, the shrinkage rate reached the maximum value of 0.95%, the bulk density reached the maximum value of 2.53 g cm^{-3}, the porosity reached the minimum value of 33.11%, and the bending strength had the highest value of 105 MPa.

Acknowledgements We would like to acknowledge the financial support provided by United Innovation Program of Shanghai Commercial Aircraft Engine (AR910), the National Natural Science Foundation of China (Nos. U1560202, 51690162), the Natural Science Foundation of Hebei Province (Grant No. A2018210123), Open Project of State Key Laboratory of Advanced Special Steel, Shanghai University (SKLASS 2017-03).

References

1. J.J. Liang, Q.H. Lin, X. Zhang, T. Jin, Y.Z. Zhou, X.F. Sun, B.G. Choi, I.S. Kim, J.H. Do, C.Y. Jo, Effects of alumina on cristobalite crystallization and properties of silica-based ceramic cores. J. Mater. Sci. Technol. **33**, 204–209 (2017)
2. W.G. Jiang, K.W. Li, J.H. Xiao, L.H. Lou, Effect of silica fiber on the mechanical and chemical behavior of alumina-based ceramic core material. J. Asian Ceram. Soc. **5**, 410–417 (2017)
3. A. Kazemi, M.A. Faghihi-Sani, H.R. Alizadeh, Investigation on cristobalite crystallization in silica-based ceramic cores for investment casting. J. Eur. Ceram. Soc. **33**, 3397–3402 (2013)
4. G.H. Gu, L.M. Cao, Development of ceramic cores for investment casting hollow blades. Found. Technol. **23**, 81–83 (2002)
5. A. Hamimah, Y. Dzul Hafez, Effect of thermal treatment on properties and micro-structure of porous alumina. Solid State Sci. Technol. **18**, 155–163 (2010)
6. Y. Yang, S.Z. Shimai, Y. Sun, M.J. Dong, H. Kamiya, S.W. Wang, Fabrication of porous Al2O3 ceramics by rapid gelation and mechanical foaming. J. Mater. Res. **28**, 2012–2016 (2013)
7. J.B. Yu, Z.G. Yang, C.J. Li, K. Deng, Z.M. Ren, Investigation on properties of the silica ceramic cores for the hollow blades prepared by the conversion of the silicon resin. Key Eng. Mater. **680**, 267–271 (2016)
8. Z.G. Yang, J.B. Yu, C.J. Li, W.D. Xuan, Z.Q. Zhang, K. Deng, Z.M. Ren, Preparing porous Si-based ceramic core using thermosetting silicon resin injection method. J. Inorg. Mater. **30**(2), 147–152 (2015)
9. Z.G. Yang, J.B. Yu, Z.M. Ren, C.J. Li, K. Deng, Fabrication and characterization of porous alumina-based ceramics using silicone resin as binder. Trans. Indian Ceram. Soc. **75**, 40–46 (2016)
10. Z.G. Yang, J.B. Yu, C.J. Li, K. Deng, Z.M. Ren, Effect of silicone resin on the properties of silica ceramic cores by heating treatment. J. Adhes. Sci. Technol. **30**(24), 2667–2677 (2016)

Effect of Reaction Temperature on CeO_2-Coated cBN Particles for Vitrified cBN Abrasive Tools

Jiang Shi, Feng He, Junlin Xie, Xiaoqing Liu and Hu Yang

Abstract Effect of reaction temperature on CeO_2-coated cBN particles for vitrified cBN abrasive tools was studied. Meanwhile, the bending strength of the abrasive tools prepared with coated and uncoated cBN abrasives was compared. Microstructure characterization confirmed that the cBN particles were successfully coated with CeO_2. Besides, CeO_2 tended to agglomerate on the cBN surface under lower reaction temperatures (55 °C), which was detrimental to the wetting of vitrified bond on the cBN surface, resulting in a decrease in the bending strength of the abrasive tools. However, as the reaction temperature increased, CeO_2 gradually spread on the surface of cBN, and the agglomeration was disappeared. The CeO_2 layer on the cBN surface gradually became dense and smooth, and the wetting state of vitrified bond on cBN was significantly improved. Thus, the holding force of the bonds on the abrasive grains was increased, which developed the bending strength of the abrasive tool. When the reaction temperature was 75 °C, T3 (AT) exhibited the best bending strength (124.54 MPa), which proved that CeO_2-coated cBN was an effective way to improve the holding force of vitrified bonds on abrasives.

J. Shi · F. He (✉) · J. Xie
State Key Laboratory of Silicate Materials for Architectures, Wuhan University of Technology, Wuhan 430070, China
e-mail: he-feng2002@163.com

J. Shi
e-mail: whutsj@126.com

J. Xie
e-mail: xjlclxy@126.com

J. Shi · F. He · J. Xie · H. Yang
School of Materials Science and Engineering, Wuhan University of Technology, Wuhan 430070, China
e-mail: 983811939@qq.com

X. Liu
Center of Materials Research and Testing, Wuhan University of Technology, Wuhan 430070, China
e-mail: lczr@whut.edu.cn

Y. Han (ed.), *Physics and Techniques of Ceramic and Polymeric Materials*, Springer Proceedings in Physics 216,
https://doi.org/10.1007/978-981-13-5947-7_2

1 Introduction

Abrasive tools, which are a kind of high-performance processing tool, are widely used in high-precision manufacturing and processing of materials. Among all types of abrasive tools, superhard abrasive tools have attracted much attention for their high processing efficiency and precision, which are significantly improved compared with traditional abrasive tools [1]. As a vital part of superhard abrasive tools, superhard abrasives and bonds can substantially affect the performances of superhard abrasive tools. To date, diamond and cubic boron nitride (cBN) are two kinds of widely used superabrasives [2, 3]. Compared with diamond, cBN presents better thermal stability [4, 5]. Also, owing to the high grinding efficiency, good grinding stability, and the highly inert to ferrous metals, cBN grinding tools show a more extensive range of applications than diamond grinding tools [6].

During the preparation of the superhard abrasive tools, bonds necessary for cBN particles are difficult to consolidate with each other spontaneously. Commonly used bonds mainly include resin bonds, metal bonds, vitrified bonds, among which vitrified bonds show the best performance and broad application prospects [7].

However, the smooth surfaces of cBN abrasives are challenging to be thoroughly wetted by vitrified bonds under high temperature [8], which results in a low bonding strength between the abrasives and bonds; thus, cBN abrasives are easy to fall off from the abrasive tools during the grinding process, and the processing efficiency and service life of the abrasive tools severely decrease. Aiming at solving this problem, serval researchers proposed methods to pretreat the surfaces of abrasives, mainly including surface activation treatment [9], surface metallization [10], and surface oxide coating. The surface treatment of the abrasives can improve the bonding strength of the superhard abrasive particles and the bonds, and avoid falloff of abrasives during the grinding process. In addition, the abrasive grains are coated with other materials to prevent thermal damages during the sintering process. However, when the surface of the abrasive is coated with metal, the wettability of vitrified bonds to the metal is not satisfactory since these bonds are composed of a plurality of oxides. So, when the surfaces of the abrasive are coated with oxides, these oxides can greatly improve the high-temperature wettability of the bonds to the abrasive by forming a solid solution with the vitrified bonds, which will contribute to upgrading the performances of vitrified bond abrasive tools. Among all kinds of oxides act as the coatings of superabrasives, SiO_2 [11], TiO_2 [12], Al_2O_3 [13], ZnO [14], TiO_2–Al_2O_3 [15] had been studied to indicate that surface oxide coating of superabrasives can achieve the desired goals. To date, researchers had found that the chemical mechanical polishing performance of SiO_2/CeO_2 was much better than SiO_2 [16, 17]. Whereas there are little publications on CeO_2-coated cBN, the cBN coated with CeO_2 will enrich the type of oxide coating on the surface of abrasives, which has high research value. In addition, the proper incorporation of CeO_2 is beneficial for improving the performances of vitrified bonds.

In this paper, micro-sized cBN powders were used to be coated on CeO_2 with the assistance of polyvinylpyrrolidone (PVP). The CeO_2-coated cBN micro-powders were prepared by hydrothermal reactions using a magnetic stirring under constant temperature. The effect of reaction temperatures on the coating state of cBN by CeO_2 was studied. In addition, the mechanical properties and microstructure of abrasive tools prepared with coated and uncoated cBN abrasives were also analyzed.

2 Materials and Methods

Figure 1 shows the schematic diagram of the experimental procedure for CeO_2-coated cBN. The experimental steps for the preparation of CeO_2-coated cBN were as follows:

(1) 1.0 g cBN and 0.1 g PVP were mixed in 70 ml ethanol; the mixed solution was marked "A." Solution "B" was prepared by dissolving 0.025 mol hexamethylenetetramine (HMT) in 20 ml distilled water; solution "C" was obtained by dissolving 0.005 mol $Ce(NO_3)_3 \cdot 6H_2O$ in 20 ml distilled water.
(2) Solution A was stirred in a magnetic stirrer for 30 min at room temperature, and then, solution B was added into solution A, and a continuous stirring for 10 min was held to obtain a thoroughly dispersed mixed solution.
(3) The above mixed solution was transferred to a constant temperature magnetic stirrer, with which solution C was gradually added; vigorously stirring was held at a constant reaction temperature for 2 h to obtain a milky white suspension. As for different reaction temperature, 55, 65, 75, and 85 °C were marked as T1, T2, T3, and T4, respectively.
(4) The suspension was slowly suction-filtered after resting for 30 min and then washed with distilled water and ethanol. Finally, the mixtures were dried at 100 °C for 12 h to obtain CeO_2-coated cBN particles.

Vitrified bond was prepared by a traditional melt-quenching method which was shown in [18]. In addition, vitrified cBN abrasive tools were prepared as follows: cBN particles, vitrified bond, and dextrin were uniformly mixed, and 2.0 g of mixed powders was pressed into 40 mm × 6 mm × 5 mm under 50 MPa for 1 min. The above shaped abrasive tools were sintered in a muffle furnace. The sintering schedule

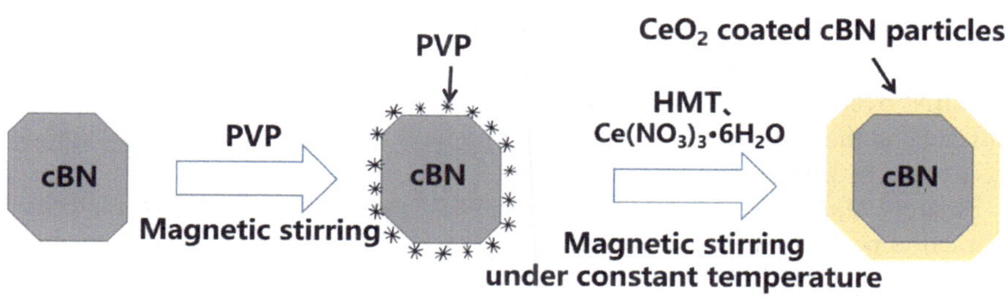

Fig. 1 Schematic diagram of the experimental procedure for CeO_2-coated cBN

was firstly heated to 500 °C at 2.5 °C/min, then raised to 860 °C at 5 °C/min, and held at this temperature for 2 h. After that, the temperature was decreased to 575 °C at 5 °C/min and kept for 1 h. Finally, the vitrified cBN abrasive tools were cooled with the furnace. Here, the ID of the abrasive tools prepared by CeO_2-coated cBN particles and the original cBN particles were named as T0 (AT) and T1 (AT)–T4 (AT), respectively.

A field-emission scanning electron microscope (FESEM; ULTRA PLUS, Zeiss) equipped with an energy-dispersive spectrometer (EDS, X-Max 50, Oxford) was used to observe the microstructure of CeO_2-coated cBN particles and the vitrified cBN abrasive tools. X-ray diffraction (XRD, D8 Advance, Bruker, Germany) with Cu Kα radiation source under a 2θ range of 10°–70° at a rate of 10°/min was utilized to identify the phase composition of samples. In addition, bending strength was conducted using a three-point test method in a universal material tester (MTS810, MTS Co, USA) at a crosshead speed of 0.5 mm/min, a fulcrum span of 25 mm, a loading speed of 9.8 $\pm$ 0.1 N/S.

3 Results and Discussions

Figure 2 is SEM image of CeO_2-coated cBN particles prepared at the various reaction temperatures. It could be seen in Fig. 2a that CeO_2 tended to agglomerate on the cBN surface under lower reaction temperatures (55 °C), which might be caused by the low reaction temperature system that cannot provide enough energy for the spread and growth of CeO_2. Thus, the thickness of the CeO_2 coating was not uniform. In addition, Fig. 2b–d indicates that the area of cBN surfaces coated with CeO_2 became larger and smoother. However, the CeO_2 coating is less dense in Fig. 2b than in Fig. 2c for the energy provided by the system under the reaction temperature of 65 °C failed to promote the complete growth and spread of CeO_2 films on cBN particles. Besides, when the reaction temperature was 85 °C, the CeO_2 coating layer became thicker, but a step-like uneven thickness distribution occurred for high temperature might promote the excessive growth of CeO_2 crystals, where CeO_2 was continuously generated on the surface of the cBN and the newly formed CeO_2 trended to cover the original thin layer, resulting in a small amount of step-like film.

For further analysis and determination of the composition of the cBN surface coating, a specific area of T2 was selected for EDS analysis. As could be seen in Fig. 3 that the area M in Fig. 3a presents the uncoated area, the EDS result shows it is cBN. Besides, area N represents the cBN surface coating area, and EDS results indicate that this area is obviously coated by a large amount of Ce; thus, the coating layer could be further determined as CeO_2.

Figure 4 shows that 2θ = 43.221°, 50.390°, and 74.055° of the characteristic diffraction peaks of T1–T4 correspond to cBN (JCPDS #74-1906). Besides, 2θ = 28.280°, 32.870°, 47.244°, 56.412°, and 76.542° are the evidence of attendance of CeO_2 (JCPDS #81-0792), indicating that the CeO_2 coexists with cBN in the prepared

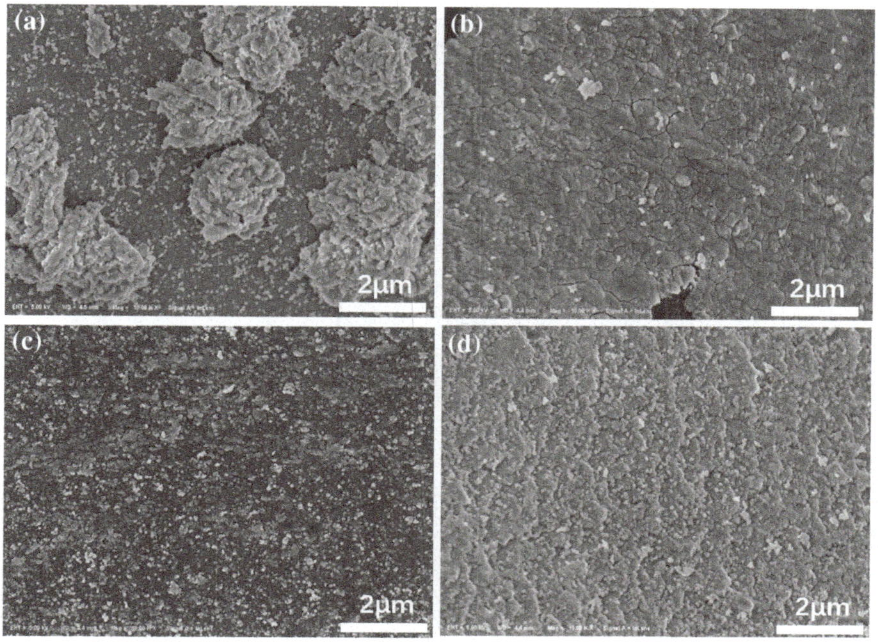

Fig. 2 SEM of CeO_2-coated cBN particles prepared at various reaction temperature **a–d** T1-T4

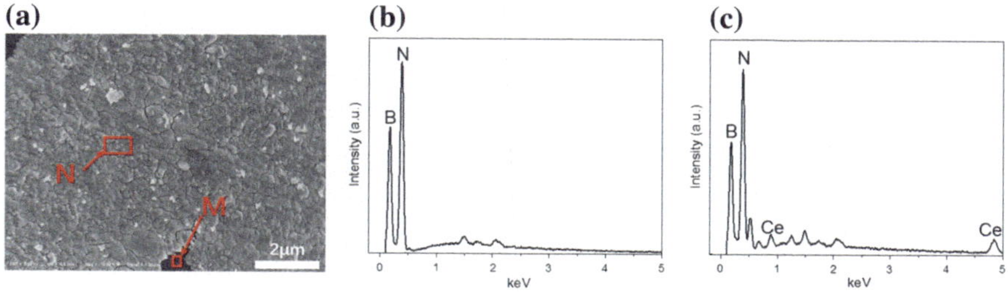

Fig. 3 SEM of (**a**) T2; EDS spectra of areas M (**b**) and N (**c**) labeled in (**a**)

samples. Combined with Figs. 2, 3, and 4, the materials coated on the surfaces of cBN are proved to be CeO_2.

In this paper, T1, T3, and uncoated cBN particles were selected as raw materials to prepare the vitrified cBN abrasive tools. As shown in Fig. 5, the bending strength of abrasive tools states the following relationship: T3(AT) > T0(AT) > T1(AT). Thus, the above relationship indicates that the presence of the smooth CeO_2 coating (T3) on cBN surfaces is beneficial to the improvement in the bending strength of abrasive tools for a well-coated CeO_2 films contributed to obtaining a good wetting and coating of cBN by the vitrified bonds, but when the CeO_2 coating is extremely uneven (T1), the coatings would adversely affect the bending strength of abrasive tools because

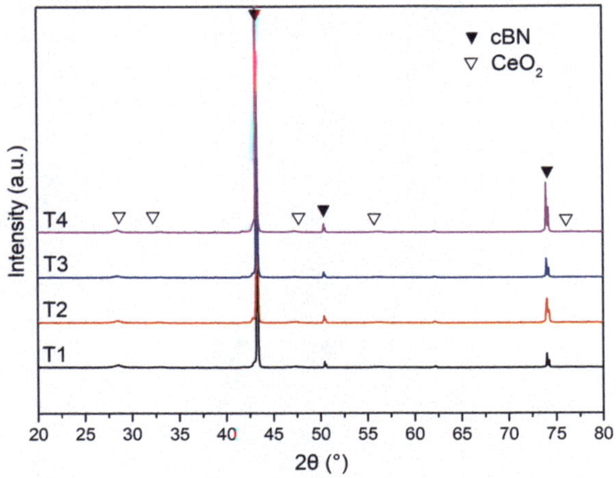

Fig. 4 XRD pattern of CeO_2-coated cBN particles

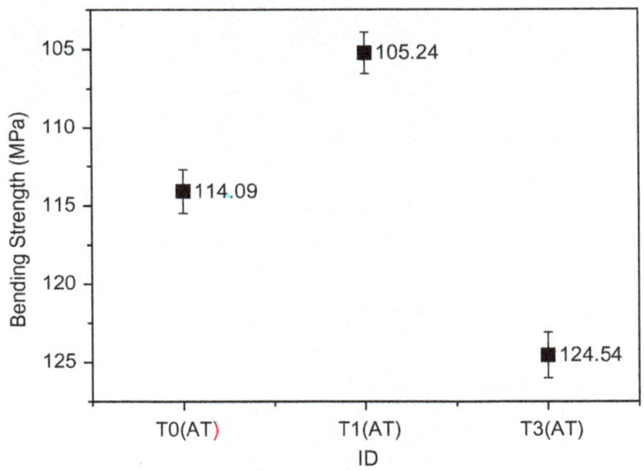

Fig. 5 Bending strength of vitrified cBN abrasive tools

the uneven CeO_2 would hinder the flow and wetting of the bond during sintering and result in a bad bonding between cBN and vitrified bonds.

Figure 6 shows the SEM image of abrasive tools. It could be seen that the number of cBN particles which failed to be wetted by vitrified bonds during high-temperature sintering among these samples revealed the following relationship: T1(AT) > T0(AT) > T3 (AT), which is consistent with the results of bending strength shown in Fig. 5; that is, the bending strength of the abrasive tools was inversely proportional to the number of cBN particles unwetted by vitrified bonds, and the worse the wettability of the cBN particles was, the lower the bending strength of the abrasive tools presented. Referring to Figs. 2, 5, and 6, it is apparent that the area of CeO_2 agglomeration on the surface of cBN under lower reaction temperature condition (55 °C) hinders the flow of glassy phase during the high-temperature sintering of vitrified cBN abrasive tools; thus, the wetting of cBN by vitrified bond is weakened, resulting in a decreased

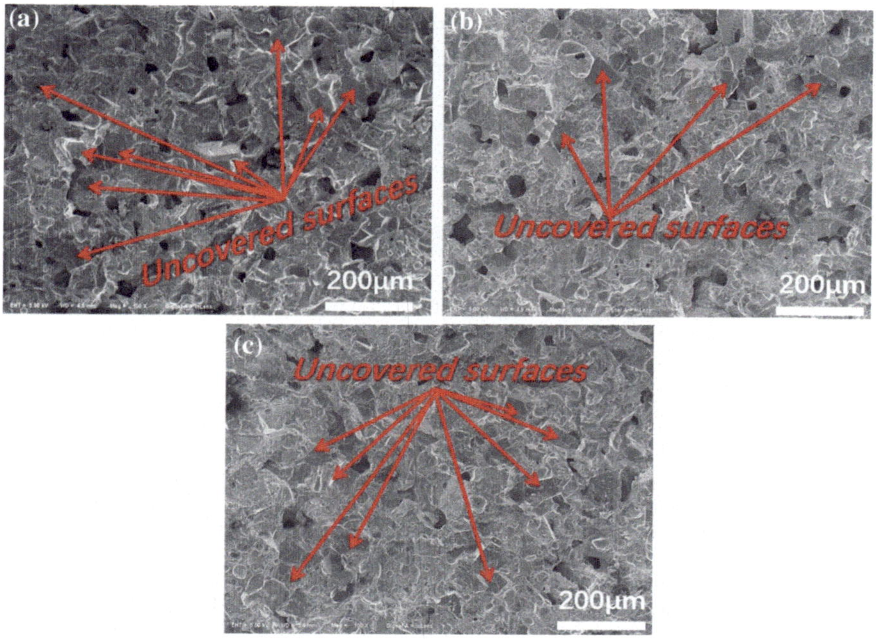

Fig. 6 SEM of abrasive tools; **a** T1 (AT); **b** T3 (AT); **c** T0 (AT)

bending strength of vitrified abrasive tools. As for T3 (AT), the smooth and dense coated CeO_2 layer facilitates the high-temperature spreading and wetting of the cBN particles by vitrified bond, which enhances the holding force of the vitrified bond on the cBN abrasives, and the bending strength of the abrasive tools also increased; similar results can be found in literatures [19, 20] which have proved that the improvement in wetting ability for vitrified bonds on superhard materials has positive impacts on the bonding strength of abrasive tools. So, it is proved that CeO_2-coated cBN is a feasible way to effectively improve the bending strength of vitrified cBN abrasive tools.

4 Conclusions

CeO_2 tended to agglomerate on the cBN surface under lower reaction temperatures (55 °C), which would hinder the flow of glassy phase to wet the cBN surfaces during the high-temperature sintering of vitrified cBN abrasive tools, resulting in a decrease in the bending strength of the abrasive tools. However, as the reaction temperature increased, CeO_2 gradually spread on the surface of cBN, and the CeO_2 layer on the cBN surface gradually became dense and smooth when the reaction temperature was 75 °C, and the wetting state of the vitrified bond on cBN was significantly improved, which improved the bending strength of the abrasive tools, and T3 (AT) exhibited the best bending strength (124.54 MPa).

References

1. J.F.G. Oliveira, E.J. Silva, C. Guo et al., Industrial challenges in grinding. CIRP Ann. Manuf. Technol. **58**(2), 663–680 (2009)
2. P.B. Mirkarimi, K.F. Mccarty, D.L. Medlin, Review of advances in cubic boron nitride film synthesis. Mat. Sci. Eng. R Rep. **21**(2), 47–100 (1997)
3. W. Zhang, I. Bello, Y. Lifshitz et al., Epitaxy on diamond by chemical vapor deposition: a route to high-quality cubic boron nitride for electronic applications. Adv. Mater. **16**(16), 1405–1408 (2004)
4. P. John, N. Polwart, C.E. Troupe et al., The oxidation of (100) textured diamond. Diam. Relat. Mater. **11**(3–6), 861–866 (2002)
5. W.J. Zhang, Y.M. Chong, I. Bello et al., Nucleation, growth and characterization of cubic boron nitride (cBN) films. J. Phys. D Appl. Phys. **40**(20), 6159–6174 (2007)
6. I.L. Pobol, A.A. Shipko, I.G. Nesteruk, Investigation of contact phenomena at cubic boron nitride-filler metal interface during electron beam brazing. Diam. Relat. Mater. **6**(8), 1067–1070 (1997)
7. F. Klocke, W. König, Appropriate conditioning strategies increase the performance capabilities of vitrified-bond cBN grinding wheels. CIRP Ann. Manuf. Technol. **44**(1), 305–310 (1995)
8. N. Yan, W. Miao, Y. Zhao et al., Effects of titania films on the oxidation resistance and dispersibility of ultrafine diamond. Mater. Lett. **141**, 92–95 (2015)
9. X. Yang, J. Bai, W. Jing et al., Strengthening of low-temperature sintered vitrified bond cBN grinding wheels by pre-oxidation of cBN abrasives. Ceram. Int. **42**(7), 9283–9286 (2016)
10. R. Chang, J. Zang, Y. Wang et al., Study of Ti-coated diamond grits prepared by spark plasma coating. Diam. Relat. Mater. **77**, 72–78 (2017)
11. D. Zhao, Z. Wang, Y. Xi et al., Preparation of silica-coated ultrafine diamond and dispersion in ceramic matrix. Mater. Lett. **113**(24), 134–137 (2013)
12. W. Miao, N. Yan, Y. Zhao et al., Synthesis and application of titania-coated ultrafine diamond abrasive particles. Ceram. Int. **42**(7), 8884–8890 (2016)
13. X. Yan, X. Li, X. Wang et al., Synthesis of nano-diamond/alumina composite by detonation method. Diam. Relat. Mater. **77**, 79–83 (2017)
14. Y. Wang, Y. Yuan, X. Cheng et al., Inhibiting the oxidation of diamond during preparing the vitrified dental grinding tools by depositing a ZnO coating using direct urea precipitation method. Mater. Sci. Eng. C **53**, 23–28 (2015)
15. W. Hu, L. Wan, X. Liu et al., Effect of TiO_2/Al_2O_3 film coated diamond abrasive particles by sol-gel technique. Appl. Surf. Sci. **257**(13), 5777–5783 (2011)
16. Y. Chen, C. Zuo, Z. Li et al., Design of ceria grafted mesoporous silica composite particles for high-efficiency and damage-free oxide chemical mechanical polishing. J. Alloy. Compd. **736**, 276–288 (2018)
17. Z. Zhang, W. Liu, J. Zhu et al., Synthesis, characterization of ceria-coated silica particles and their chemical mechanical polishing performance on glass substrate. Appl. Surf. Sci. **257**(5), 1750–1755 (2010)
18. J. Shi, F. He, J. Xie et al., Kinetic analysis of crystallization in Li_2O-Al_2O_3-SiO_2-B_2O_3-BaO glass-ceramics. J. Non-Cryst. Solids **491**, 106–113 (2018)
19. Y. Yin, J. Fang, Z. Qin et al., Toughening effect and mechanical behaviour of SiC@SiO_2 whisker-reinforced vitrified diamond composites. Int. J. Refract. Metal Hard Mater. **71**, 190–197 (2018)
20. D. Song, L. Wan, X. Liu et al., Influence of diamond particles coated with TiO_2 film on wettability of vitrified bond and transverse rupture strength (TRS) of vitrified bond composites. J. Mater. Eng. Perform. **25**(6), 2282–2287 (2016)

Bifunctional Roles of Dialdehyde Cellulose Nanocrystals in Reinforcing and Cross-Linking Electrospun Chitosan Nanofibrous Membranes

Jiaming Liu, Zongyi Qin, Miao Cheng, Fanxin Zeng, Shuo Hu and Youwei Zhang

Abstract The composite nanofibrous membranes were electrospun by adding separately dialdehyde cellulose nanocrystals (DACNCs) and cellulose nanocrystals (CNCs) as the nanofillers into chitosan (CS) matrix. The morphology, chemical structure, thermal stability and mechanical properties of as-prepared composite membranes were investigated by scanning electron microscopy (SEM), Fourier transform infrared spectroscopy (FT-IR), thermogravimetric (TGA) and mechanical test, respectively. It is found that significant enhancements in the mechanical properties and thermal stability of CS nanofibrous membranes could be achieved by introducing a small amount of cellulose-based nanofillers. Moreover, greater improvements could be found in the presence of the DACNCs, which could be attributed to the bifunctional roles of DACNCs played simultaneously as the reinforcing and cross-linking reagents for CS nanofibrous membranes.

1 Introduction

Chitosan (CS) has exhibited great potential applications in wound dressing, wound healing, drug delivery systems and so on due to its biologically renewable, biodegradable, nonantigenic and biocompatible characteristics [1]. Especially, electrospun CS nanofibrous membranes with three-dimensional porous structures have attracted more attentions as possible biomedical materials. Ohkawa et al. have successfully prepared the electrospinning solution of CS by using trifluoroacetic acid (TFA) as an appropriate solvent to effectively destroy the rigid interactions among the CS molecules through strong interaction between the amino groups of the CS with

J. Liu · Z. Qin (✉) · M. Cheng · F. Zeng · S. Hu
State Key Laboratory for Modification of Chemical Fibers and Polymer Materials, Donghua University, Shanghai 201620, China
e-mail: phqin@dhu.edu.cn

J. Liu · M. Cheng · F. Zeng · S. Hu · Y. Zhang (✉)
College of Material Science and Engineering, Donghua University, Shanghai 201620, China
e-mail: zyw@dhu.edu.cn

Y. Han (ed.), *Physics and Techniques of Ceramic and Polymeric Materials*,
Springer Proceedings in Physics 216,
https://doi.org/10.1007/978-981-13-5947-7_3

TFA molecules [2]. However, the application of neat CS nanofibrous membrane has been limited because of their poor mechanical properties. The incorporation of the nanofillers into the CS matrix is one of the most effective methods for enhancing thermal and mechanical properties of nanofibrous composite membranes. It has been demonstrated that cellulose nanocrystals (CNCs) could be used as efficient nanofillers to improve the mechanical properties of electrospun nanofibrous membranes of various biodegradable polymers [3]. Therefore, it is expected that the mechanical properties of CS nanofibrous membrane could be enhanced by selecting the rod-like CNCs as the nanofillers due to their similar chemical structures with the CS as well as high-specific strength and modulus. However, the CNCs prepared through acid hydrolysis had abundant surface hydroxyl groups, which usually would result in their poor dispersion in polymeric matrix. Fortunately, hydroxyl groups of these CNCs could be changed into two aldehyde groups per glucose unit through specific cleavage of the C2–C3 bond of the glucopyranoside ring by periodate oxidation, that is to say, dialdehyde celluloses nanocrystals (DACNCs) could be obtained [4]. In this work, two kinds of CNCs including the CNCs with abundant surface hydroxyl groups and two aldehyde groups were introduced into electrospun CS nanofibrous membrane, respectively. It is worth to point out that DACNCs could serve as not only reinforcing nanofillers but also cross-linking reagents, and thus, more attentions would be paid to the bifunctional effects of the DACNCs on the microstructure and properties of the composite nanofibrous membranes.

2 Experimental

2.1 Materials

Microcrystalline cellulose (MCC), sodium periodate ($NaIO_4$), hydrochloric acid (HCl) and trifluoroacetic acid (TFA) were obtained from Sinopharm Chemical Reagent Co., Ltd. Chitosan powder with an average molecular weight of 250,000 and 75% degree of deacetylation was purchased from Yuhuan Ocean Biochemistry.

2.2 Preparation

The preparation of the CNCs was according to the method reported previously [5]. Briefly, 1 g MCC as the starting materials were added into 6 M HCl solution in hydrothermal kettle and reacted at 110 °C for 3 h. Subsequently, the as-prepared CNCs were treated in 0.3 M $NaIO_4$ aqueous solution at 45 °C for 30 min for the preparation of DACNCs.

The electrospinning solution of 4.0 wt% CS was firstly prepared by dissolving the CS powder was dissolved into TFA. Then, the CNCs and DACNCs with the fraction of

1.0 wt% were added into the above solution, respectively. Then, the electrospinning was carried out by applying a potential of 25 kV to the syringe needle with the diameter 12.25 mm, and the feeding rate of the spinning solution was controlled at 0.8 mL/h. The electrospun nanofibrous membrane was collected on aluminum foil which was placed at a distance of 15 cm from the syringe tip.

2.3 *Characterization*

The morphologies of the CNCs and DACNCs as well as neat CS and composite nanofibrous membranes were observed by a HITACHI S-4800 field emission scanning electron microscopy (FE-SEM). The chemical structures, thermal and mechanical properties were characterized by a Nicolet 8700 Fourier transform infrared (FT-IR) spectrometer, NETZSCH TG 209 F1 thermogravimetric analyzer (TGA) and electronic universal testing machine (Kexin WDW3020), respectively.

3 Results and Discussion

Figure 1 shows the SEM images and FT-IR spectra for the CNCs and DACNCs. It is noted that both two kinds of the nanocrystals possessed a rod-like structure, and only slight difference in the diameter could be observed. More detailedly, the average length and diameter of the CNCs and DACNCs were 242 ± 6 and 24 ± 2 nm as well as 234 ± 10 and 20 ± 4 nm, respectively. After the oxidation treatment, the diameter of nanocrystals remained almost unchanged, implying no seriously damage of the rigid structure of the nanocrystals in relatively short reaction time of 30 min. All the peaks of cellulose appeared in the infrared spectra of the CNCs and DACNCs. However, compared to those for the CNCs, a new peak at 1725 cm^{-1} for the C=O vibration could be found, indicating the formation of aldehyde groups on the surface of the nanocrystals by treating CNCs in $NaIO_4$ solution [6].

Figure 2 displays the SEM images and FT-IR spectra of neat CS, CS-CNCs and CS-DACNCs nanofibrous membranes. It is found that neat CS nanofibers showed smooth surface with fewer defects, and the average diameter of the nanofibers was about 255 ± 51 nm. For the composite nanofibers, the average diameter increased to 306 ± 102 nm for the CNCs and 408 ± 204 nm for DACNCs, respectively. The larger dimension for DACNCs could be attributed to the chemical cross-linking between CS and DACNCs which would limit the jet elongation. Moreover, stronger infrared peaks at about 1630 cm^{-1} corresponding to C=N vibration and the disappearance of the peaks at 1730 cm^{-1} to C=O vibration further indicated the occurrence of the cross-linking reaction among the residual carbonyl groups in DACNCs and the amino groups in CS [7].

Figure 3 presents the TGA curves (a) and corresponding thermal parameters (b) for neat CS, CS-CNCs and CS-DACNCs nanofibrous membranes. It is found that

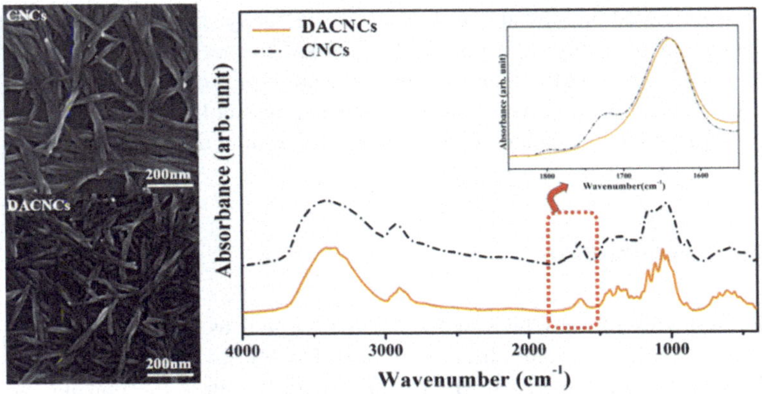

Fig. 1 SEM images and FT-IR spectra of the CNCs and DACNCs

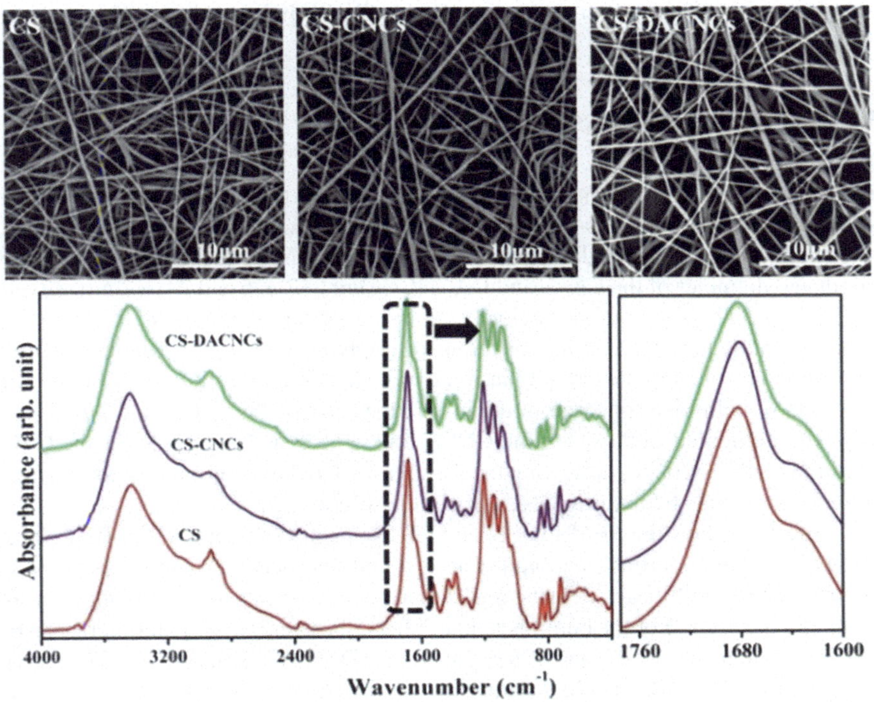

Fig. 2 SEM images and FT-IR spectra of neat CS and resulting composite nanofibrous membranes with 1.0 wt% CNCs and DACNCs

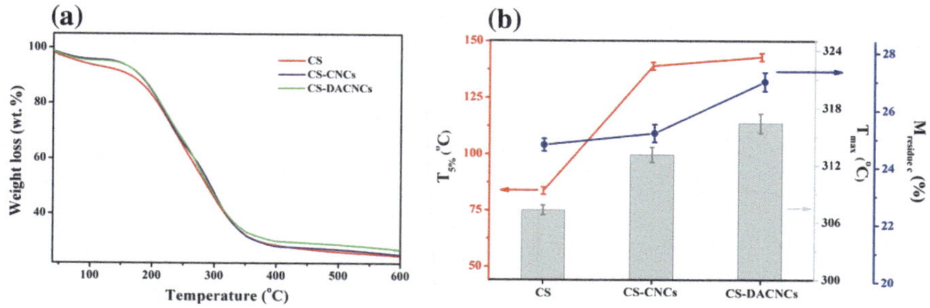

Fig. 3 TGA curves (**a**) and thermal parameters (**b**) of nanofibrous membranes

the thermal degradation for all the membranes took place in only one step under an inert atmosphere as shown in Fig. 3a. The removal of water absorbed on the surface of the nanocrystals occurred firstly before 100 °C. For neat CS nanofibrous membranes, the decomposition appeared mainly within the temperature ranges of 170–345 °C, which was associated with thermal scission of the polymer backbone [8]. As given in Fig. 3b, the maximum decomposition temperatures (T_{max}), the mass loses 5% of temperature ($T_{5\%}$) and the amount of residue carbon ($M_{residue\ c}$) of the composites membranes became higher compared with those for neat CS membranes. Although there were no obvious differences in $T_{5\%}$ between two kinds of composite membranes, more carbon residue and the higher T_{max} could be found in the presence of the DACNCs.

Figure 4 shows the stress-strain curves and corresponding mechanical properties of neat CS, CS-CNCs and CS-DACNCs nanofibrous membranes. It can be observed in Fig. 4a that the stress-strain curves for CS nanofibrous membranes would shift toward higher values of stress when the cellulose-based nanofillers were added into CS matrix, indicating the reinforcement of the nanofillers. Furthermore, as illustrated in Fig. 4b, the tensile strength of the composite nanofibrous membranes increased from 20.5 ± 0.35 MPa for neat CS membrane to 30.2 ± 0.54 and 35.8 ± 0.62 MPa, respectively, meanwhile the composite membranes showed 13 and 59% improvement in elastic modulus compared with the CS nanofibrous membranes. It is clear that the addition of nanofillers could reduce the loading force on CS matrix through the stress transfer from CS to nanofillers due to the formation of strong hydrogen bonding interaction in the composites [9]. More importantly, as plotted in Fig. 4b, the DACNCs could exhibit stronger reinforcing effect for CS nanofibrous membranes compared with the CNCs, which could be contributed to bifunctional roles of the DACNCs played simultaneously as reinforcing nanofillers and cross-linking regent [10].

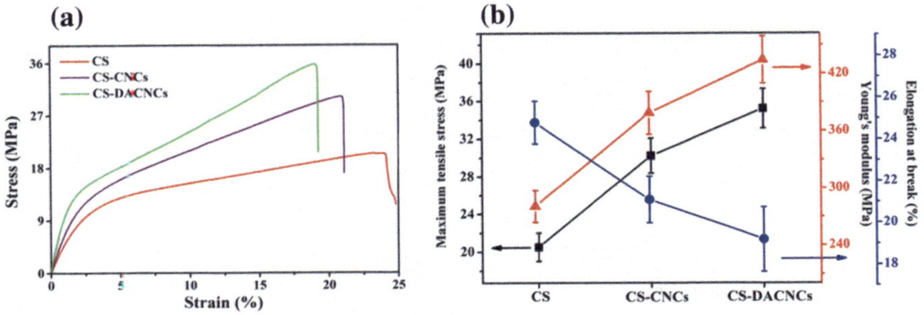

Fig. 4 Typical tensile stress-strain curves (**a**) and mechanical properties (**b**) of nanofibrous membranes

4 Conclusions

Two kind of CS-based composite nanofibrous membranes were electrospun by adding the CNCs with different surface groups as the nanofillers. Compared with neat CS nanofibers, the composite nanofibers exhibited larger diameter with wider distribution especially in the presence of the DACNCs. The infrared spectra analysis indicated the strong hydrogen bonding between the CS and cellulose-based nanofillers appeared in the composite nanofibrous membranes, and moreover the cross-linking reaction occurred when the DACNCs was introduced into CS matrix. Although only slight difference in the $T_{5\%}$ for all the membranes could be found, the composite membranes containing the DACNCs exhibited the largest T_{max} and most carbon residue. Furthermore, compared with neat CS nanofibrous membrane, all the composite membranes exhibited significant improvements in the thermal stability and mechanical properties. The tensile strength of the composite nanofibrous membranes increased from 20.5 ± 0.35 MPa for neat CS membrane to 30.2 ± 0.54 and 35.8 ± 0.62 MPa, respectively, accompanying with 13 and 59% improvements in elastic modulus. Obviously, the composite membranes in the presence of the CNCs exhibited the highest mechanical properties among all the electrospun nanofibrous membranes, which could be attributed to simultaneously existence of the hydrogen bonding and cross-linking interaction between the CS and the DACNCs.

Acknowledgements This work has been financially supported by Program for Innovative Research Team in University of Ministry of Education of China (IRT_16R13). Prof. Qin also appreciates the Talent Support Program for the Western Shandong Economy Uplift Belt (2017).

References

1. V.K. Thakur, S.I. Voicu, Recent advances in cellulose and chitosan based membranes for water purification: a concise review. Carbohyd. Polym. **146**, 148–165 (2016)
2. K. Ohkawa, D. Cha, H. Kim, A. Nishida, H. Yamamoto, Electrospinning of chitosan. Macromol. Rapid Commun. **25**, 1600–1605 (2004)

3. N. Naseri, A.P. Mathew, L. Girandon, Porous electrospun nanocomposite mats based on chitosan-cellulose nanocrystals for wound dressing: effect of surface characteristics of nanocrystals. Cellulose **22**(1), 521–534 (2015)
4. P.R. Sharma, A.J. Varma, Thermal stability of cellulose and their nanoparticles: effect of incremental increases in carboxyl and aldehyde groups. Carbohydr. Polym. **114**, 339–343 (2014)
5. H. Yu, Z. Qin, B. Liang, N. Liu, Z. Zhou, L. Chen, Facile extraction of thermally stable cellulose nanocrystals with a high yield of 93% through hydrochloric acid hydrolysis under hydrothermal conditions. J. Mater. Chem. A **1**(12), 3938–3944 (2013)
6. V. Verma, P. Verma, P. Ray, A.R. Ray, 2, 3-dihydrazone cellulose: prospective material for tissue engineering scaffolds. Mater. Sci. Eng. C **28**(8), 1441–1447 (2008)
7. B. Sun, Q. Hou, Z. Liu, Y. Ni, Sodium periodate oxidation of cellulose nanocrystal and its application as a paper wet strength additive. Cellulose **22**(2), 1135–1146 (2015)
8. A. Khan, R.A. Khan, S. Salmieri, C. Le Tien, B. Riedl, J. Bouchard, M. Lacroix, Mechanical and barrier properties of nanocrystalline cellulose reinforced chitosan based nanocomposite films. Carbohyd. Polym. **90**(4), 1601–1608 (2012)
9. U.T.M. Sampath, Y.C. Ching, C.H. Chuah, R. Singh, P.C. Lin, Preparation and characterization of nanocellulose reinforced semi-interpenetrating polymer network of chitosan hydrogel. Cellulose **24**(5), 2215–2228 (2017)
10. X. Tian, D. Yan, Q. Lu, X. Jiang, Cationic surface modification of nanocrystalline cellulose as reinforcements for preparation of the chitosan-based nanocomposite films. Cellulose **24**(1), 163–174 (2017)

Fabrication of Porous SiO_2 Nanofibers by Electrospinning with the Anti-solvent Process

Zhaowei Liu, Yufei Tang, Wenhao Chen and Kang Zhao

Abstract Inorganic nanofibers are widely used in wastewater treatment for its excellent chemical corrosion resistance and large specific surface area. In order to meet the requirements for high adsorption effect, the specific surface area of inorganic nanofibers needs to be further improved. In this study, internal dense porous SiO_2 nanofibers were prepared using the electrospinning technique combined with the anti-solvent method. The effects of receiving distances on the composite nanofibers' morphology and diameter were investigated. The diameter distribution and pore size distribution of porous nanofibers were statistically characterized. Results indicated that internal dense porous SiO_2 nanofibers could be prepared by electrospinning combined with the anti-solvent method. With the increase of the receiving distance from 10 to 17 cm, the average diameter of the PS/SiO_2 composite nanofibers decreased from 1264.4 to 766.8 nm. When the receiving distance was 10 cm, the average diameter of the nanofibers was 906.4 nm and the average surface hole was 181.7 nm after calcination. And the BET surface area of the nanofibers was 78.94 m^2/g. Such porous nanofibers have potential applications in adsorbents and oil–water separation.

1 Introduction

Water pollution and shortage of freshwater resources become increasingly serious, and sewage treatment technology has attracted more and more attention. Nanofibers have been gaining considerable attention because of its high porosity, large specific surface area, and excellent filtration effect [1]. In particular, inorganic nanofiber membrane is widely used in wastewater treatment [2], oil–water separation [3], ion adsorption [4] and energy storage system [5], for its excellent properties, such as high temperature resistance, excellent chemical corrosion resistance, and no secondary pollution [6]. However, the specific surface area of the inorganic nanofibers can still be further improved to meet the requirements for the higher adsorption

Z. Liu · Y. Tang (✉) · W. Chen · K. Zhao
Xi'an University of Technology, Xi'an 710048, China
e-mail: yftang@xaut.edu.cn

Y. Han (ed.), *Physics and Techniques of Ceramic and Polymeric Materials*,
Springer Proceedings in Physics 216,
https://doi.org/10.1007/978-981-13-5947-7_4

effect and wider application [7]. Preparing nanofibers with extremely small pore size and highly porous structure is an effective way to solve this problem [8]. Electrospinning is a relatively easy, fast, and cost-effective technology of producing continuous one-dimensional nanofibers [9]. And porous nanofibers can be effectively prepared by adjusting the electrospinning parameters [10] or the nature of solution components [11], which has gradually become a hot topic in the research field of nanomaterials. For instance, Yu et al. [12] reported the controllable preparation of mesoporous V_2O_5 nanofibers by electrospinning a polymer solution containing the sol–gel precursors, followed by annealing in air. Zhang et al. [13] prepared porous SiO_2 nanofibers through electrospinning PAN/PMMA/SiO_2 composites and subsequent sintering treatment to remove PMMA component. Pore size and pore volume of porous silica fibers were controllable by adjusting the content of SiO_2 nanospheres and the weight ratio of PAN/PMMA. In this case, nanofibers with porous structure inside and outside are easily and efficiently prepared. However, to best of our knowledge, the strength of the nanofibers was reduced due to the internal porous structure [14], and the adsorption effect on the wastewater was not improved [15].

Based on these researches, in this study, internal dense SiO_2 nanofibers and superficial porous structure were prepared using the electrospinning technique combined with the anti-solvent method. The morphology of the composite nanofibers prepared at the different receiving distances was investigated. The diameter distribution and pore size distribution of porous nanofibers were statistically characterized. Finally, the mechanism of porous nanofibers formation was analyzed.

2 Experimental Procedure

2.1 *Materials and Methods*

Ethyl silicate (TEOS, AR, Tianjin Kermel Chemical Reagent Co., Ltd., Tianjin, China) was used as Si precursors, respectively. *N*,*N*-dimethylformamide (AR, Tianjin Kermel Chemical Reagent Co., Ltd., Tianjin, China) was used as solvents. Phosphoric acid (AR, Tianjin Yaohua chemical reagents Co., Ltd., Tianjin, China) was used as a hydrolysant for TEOS. Polystyrene (PS, Mw = 1,300,000) was purchased from Sigma–Aldrich (USA) and used as spinning agent.

N,*N*-dimethyl formamide (DMF) solution containing 20 wt% polystyrene (PS) was prepared with strong stirring for more than 6 h. Then equal volume of Ethyl silicate (TEOS) and three drops of phosphoric acid were added into the solution and kept stirring for 24 h to get a homogeneous spinning solution. The SiO_2 spinning solution was loaded into a micropump, and PS/SiO_2 composite nanofibers were obtained after electrospinning. A voltage of 24 kV direct current (DC) was applied between the cathode (Ethanol-containing ice–water mixture) and the anode (syringe tip), and the distance between the two electrodes was 10 cm, 14 cm, or 17 cm. After a period of electrospinning, the as-collected composite nanofibers were placed into a muffle furnace and calcined (0.5 °C/min) in air at 600 °C for 2 h to form porous SiO_2 nanofibers.

2.2 *Characterization*

Morphologies of PS/SiO_2 composite nanofibers and porous SiO_2 nanofibers were observed by scanning electron microscopy (SEM, JSM 6390A; JEOL, Tokyo, Japan). The diameter, the pore size, and the surface hole of the nanofibers were determined by Adobe Photoshop CS6 software from the SEM images; herein, 50 nanofibers, 50 pores, and 100 holes at least were measured in this manner to obtain the distribution and the average value. The porous structure of SiO_2 nanofibers was characterized by transmission electron microscopy (TEM, JEM3010, JEOL, Japan). The specific surface area of porous SiO_2 nanofibers was characterized by using the Brunauer–Emmett–Teller (BET) nitrogen adsorption method (Gemini VII 2390).

3 Results and Discussion

Figure 1 shows the morphologies and the diameter distributions of PS/SiO_2 composite nanofibers obtained by electrospinning with the anti-solvent method at the different receiving distances. The surface of composite nanofibers produced at 10 cm is porous, and the diameter size of nanofibers is nonuniform. When the receiving distance is 14 cm, the diameter of the fibers becomes smaller and nanoholes decrease rapidly on the fiber surface when the receiving distance is 17 cm. The nanoholes structure is easily formed on the fiber surface at the shortest receiving distance due to the non-volatile solvent effect when the jet enters the receiving vessel, and the polymer is rapidly precipitated at this point. When the receiving distance increases, the solvent in the jet is basically volatilized and the fiber is basically formed before entering the receiving vessel, so the number of nanoholes decreases and even no hole structure is formed on the surface.

With the increase of the receiving distance, the average diameter of the PS/SiO_2 composite nanofibers decreases from 1264.4 to 766.8 nm. And the diameter of the fibers becomes small and uniform (green line shown in images of Fig. 1d). This finding could be due to the increase of the jet time under the high-voltage electric field stretching as the receiving distance increases. And at the same time, the smaller diameter results in a larger area of contact between the jet and the air, which could increase solidification rate of the jet to form the fibers. This is also a reason for the reduction in the number of nanoholes on the fiber surface.

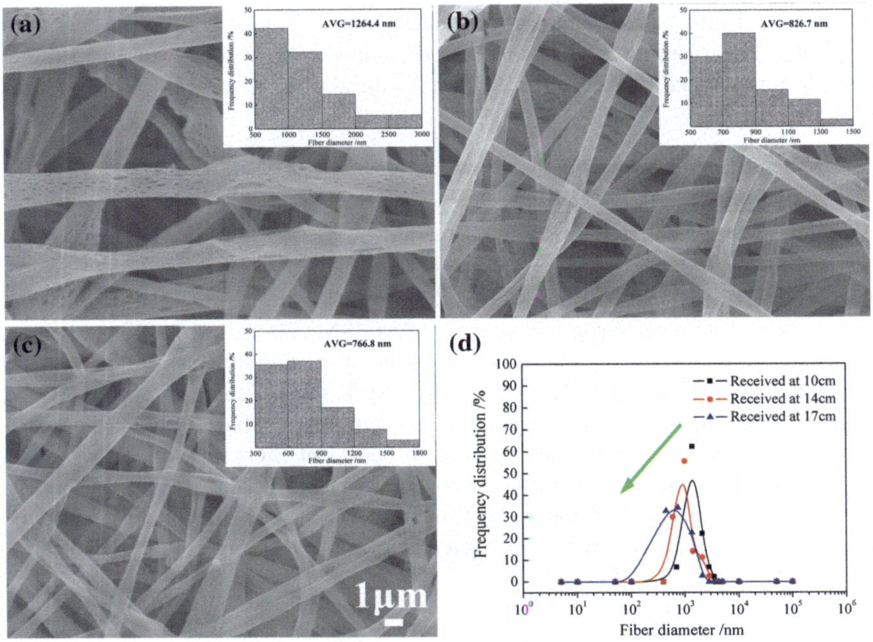

Fig. 1 Morphologies and diameter distributions of PS/SiO_2 composite nanofibers obtained by electrospinning with the anti-solvent method at the different receiving distances: **a** 10 cm, **b** 13 cm, **c** 17 cm, and **d** the diameter distribution trend analysis

Thermochemical conversion behavior of PS/SiO_2 composite nanofibers during pyrolysis was investigated using thermogravimetric curves, TG and DSC. Figure 2 shows TG–DSC curves for PS/SiO_2 composite nanofibers and the morphologies of porous SiO_2 nanofibers obtained by electrospinning with the anti-solvent method at 10 cm. It can be seen from Fig. 2a that the maximum mass loss rate is 80% when the temperatures was 500 °C. The nanoholes structure becomes apparent, and the number of nanoholes increases on the fiber surface after calcination [16]. Some adhesions are observed at Fig. 2b. When the temperature rises during the calcination, the organic polymers would soften gradually and bond together between two cross nanofibers. In addition, the depth and the number of the nanoholes on the fiber surface increase. This is attributed to the organic polymers disappear after sintering. The inside of the fibers is dense (Fig. 2c). It can be also clearly seen from Fig. 2d that the edge of the TEM image is transparent, and the color of the center is heavy, which means that the fiber surface is a porous structure while the central part is a solid structure.

The diameter, the pore size, the surface hole, and the BET surface area of the porous SiO_2 nanofibers (Fig. 2b) were characterized, as shown in Fig. 3. The average diameter of the nanofibers decreases from 1264.4 to 906.4 nm after calcination. And

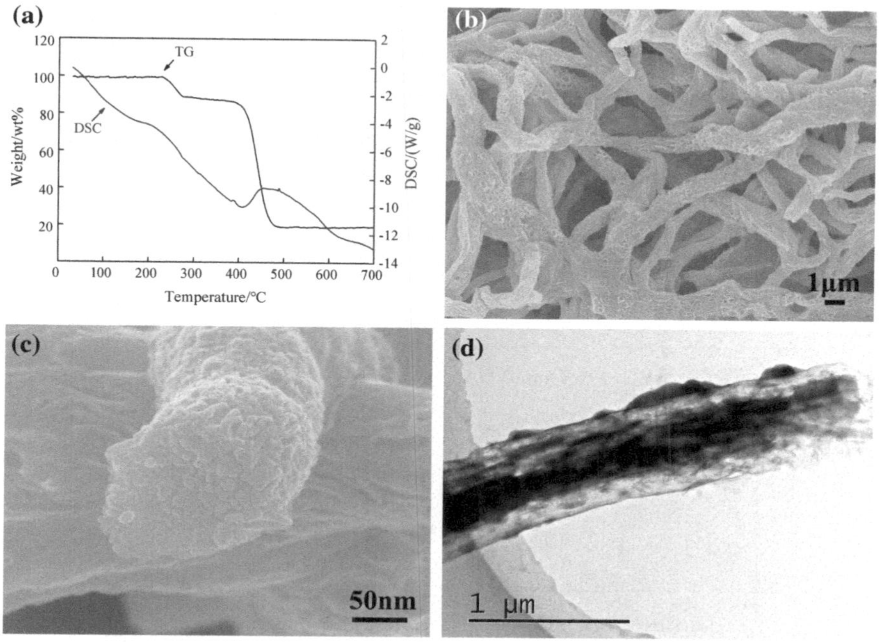

Fig. 2 **a** TG-DSC curves for PS/SiO_2 composite nanofibers; **b** morphologies of porous SiO_2 nanofibers; **c** cross-sectional morphologies of single SiO_2 fiber; **d** TEM micrograph of single SiO_2 fiber

the average pore size is 996.8 nm and the average surface hole is 181.7 nm. The nanofibers are observed with surface holes concentrated in the range of 100–200 nm, and the BET surface area is 78.94 m^2/g. As the receiving distance increases, the specific surface area decreases first and then increases, which is due to the combined effect of decrease in the pore structure and the diameter of the nanofibers.

Figure 4 is a schematic showing the formation of the porous SiO_2 nanofibers obtained by electrospinning with the anti-solvent method. During the electrospinning process, stretching of the solution occurs when the high voltage is applied [17]. Due to the stretching of the solution, jet formation starts from the Taylor cone, and the nanofibers are collected at the receiving vessel [18]. During the process of fiber formation, the solvent (DMF) starts to evaporate. With the evaporation of the DMF, it spreads from the center to the surface. When the jet is immersed into the receiving vessel, polystyrene (PS) is precipitated sharply because its solubility significantly decreases from the DMF to the water [19]. Since water is miscible with DMF, they mix well with each other on the nanofiber surface. After drying and calcination, the porous SiO_2 nanofibers with internal dense are prepared.

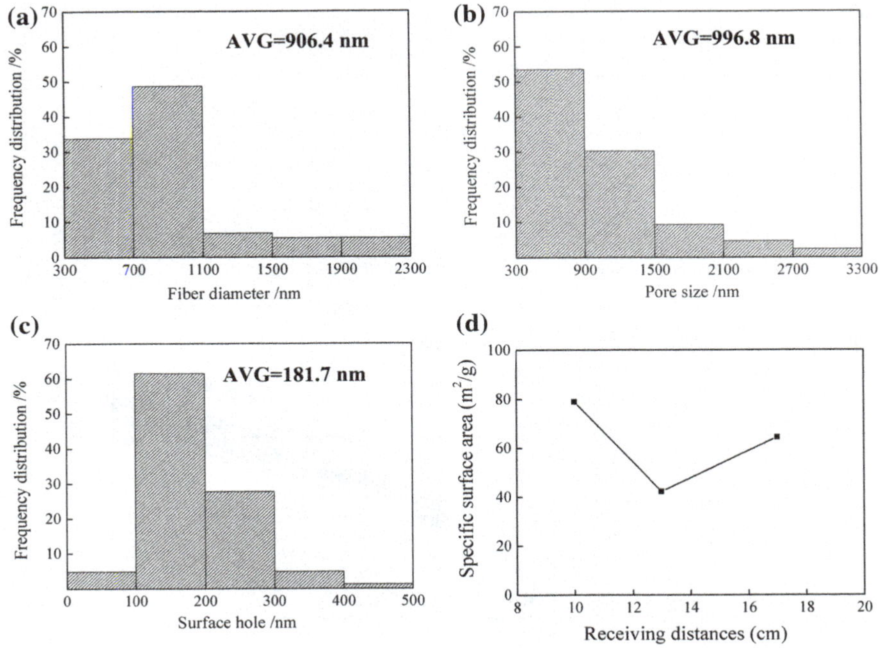

Fig. 3 **a** Diameter distributions; **b** pore size distributions; **c** surface hole distributions; and **d** BET surface area of porous SiO_2 nanofibers obtained by electrospinning with the anti-solvent method

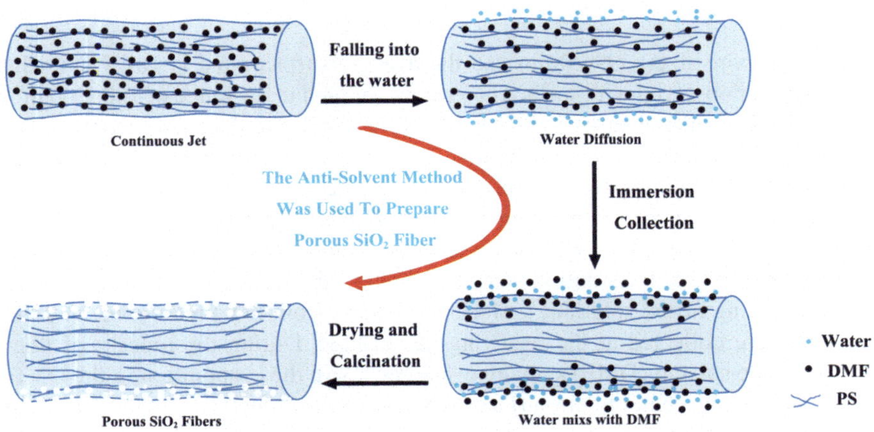

Fig. 4 Schematic showing the formation of the porous SiO_2 nanofibers obtained by electrospinning with the anti-solvent method

4 Summary

Internal dense SiO_2 nanofibers and superficial porous structure were prepared by electrospinning combined with the anti-solvent method. With the increase of the receiving distance from 10 to 17 cm, the average diameter of the PS/SiO_2 composite nanofibers decreased from 1264.4 to 766.8 nm. When the receiving distance was 10 cm, the average diameter of the nanofibers was 906.4 nm after calcination, the average pore size was 996.8 nm and the average surface hole was 181.7 nm. The surface holes were concentrated in the range of 100–200 nm, and the BET surface area of the nanofibers was 78.94 m^2/g. For its excellent properties, the internal dense porous nanofibers will be potentially used in wastewater treatment, oil–water separation, and ion adsorption in the future.

Acknowledgements The authors would like to acknowledge the support from the National Natural Science Foundation of China (No. 51572217), the China Postdoctoral Science Foundation (No. 2015M582689) and the Postdoctoral Science Foundation of Shaanxi Province (No. 2016BSHEDZZ03).

References

1. S. Barth, Synthesis and applications of one-dimensional semiconductors. Prog. Mater Sci. **55**(6), 563–627 (2010)
2. Y. Tang, Positively charged and flexible $SiO_2@ZrO_2$ nanofibrous membranes and their application in adsorption and separation. RSC Adv. **8**(23), 13018–13025 (2018)
3. J. Gao, Facile preparation of hybrid microspheres for super-hydrophobic coating and oil-water separation. Chem. Eng. J. **326**, 443–453 (2017)
4. W. Qin, Fabrication of porous chitosan membranes composed of nanofibers by low temperature thermally induced phase separation, and their adsorption behavior for Cu^{2+}. Carbohyd. Polym. **178**, 338–346 (2017)
5. H. Wang, Synthesis of SnO2 versus Sn crystals within N-doped porous carbon nanofibers via electrospinning towards high-performance lithium ion batteries. Nanoscale **8**(14), 7595–7603 (2016)
6. A. Baji, One-dimensional multiferroic bismuth ferrite fibers obtained by electrospinning techniques. Nanotechnology **22**(23), 235702 (2011)
7. A.E. Danks, The evolution of 'sol–gel' chemistry as a technique for materials synthesis. Mater. Horiz. **3**(2), 91–112 (2016)
8. A. Raza, In situ cross-linked superwetting nanofibrous membranes for ultrafast oil–water separation. J. Mater. Chem. A **2**(26), 10137–10145 (2014)
9. A.K. An, PDMS/PVDF hybrid electrospun membrane with superhydrophobic property and drop impact dynamics for dyeing wastewater treatment using membrane distillation. J. Membr. Sci. **525**, 57–67 (2017)
10. H. Fashandi, Pore formation in polystyrene fiber by superimposing temperature and relative humidity of electrospinning atmosphere. Polymer **53**(25), 5832–5849 (2012)
11. J.Y. Park, Preparation of electrospun porous ethyl cellulose fiber by THF/DMAc binary solvent system. J. Ind. Eng. Chem. **13**(6), 1002–1008 (2007)
12. D. Yu, Mesoporous vanadium pentoxide nanofibers with significantly enhanced Li-ion storage properties by electrospinning. Energy Environ. Sci. **4**(3), 858–861 (2011)

13. R. Zhang, Nanoporous fibers built with carbon-bound SiO_2 nanospheres via electrospinning and calcination. Mater. Des. **130**, 231–238 (2017)
14. X. Liao, Effect of porous structure on mechanical properties of C/PLA/nano-HA composites scaffold. Frontier Symp. China Postductors Mater. Sci., 748–751 (2009)
15. G. Arran, The effect of pore structure on the CO_2 adsorption efficiency of polyamine impregnated porous carbons. Microporous Mesoporous Mater. **208**, 129–139 (2015)
16. W. Luo, Electrospun porous $ZnCo_2O_4$ nanotubes as a high-performance anode material for lithium-ion batteries. J. Mater. Chem. **22**(18), 8916–8921 (2012)
17. M. Bognitzki, Nanostructured fibers via electrospinning. Adv. Mater. **13**(1), 70–72 (2001)
18. S. Megelski, Micro-and nanostructured surface morphology on electrospun polymer fibers. Macromolecules **35**(22), 8456–8466 (2002)
19. H.S. Bae, Fabrication of highly porous PMMA electrospun fibers and their application in the removal of phenol and iodine. J. Polym. Res. **20**(7), 1–7 (2013)

The Mechanical Property and Crystalline Structure of Novel High-Strength Polyamide Fibers

Weinan Pan, Jiaguang Sheng, Hengxue Xiang, Linggen Kong, Gongxun Zhai, Chi Ma, Mingda Ni and Meifang Zhu

Abstract A kind of novel high-strength copolyamide 6/66 fibers prepared by an effective and simple approach is reported in this paper. Firstly, PA6/66 copolyamide was obtained by polymerization of caprolactam and nylon 66 salt, and then PA6/66 fibers were prepared after melt-spinning and drawing process. The effects of draft ratio and drawing temperature on the mechanical property and crystalline structure of PA6/66 fibers were investigated. The results revealed that a complete disappearance of γ crystal occurred in PA6/66 fibers after drawing. When the draft ratio was 4.0, the breaking strength of PA6/66 fibers was 4.57 cN/dtex, which was easy to reach the standard of polyamide fiber superior product (4.0 cN/dtex). Furthermore, the highest breaking strength of PA6/66 fibers could be achieved (5.12 cN/dtex) when the draft ratio was 4.5. As the draft ratio increased, the crystallinity of PA6/66 fiber also increased from 44.5 to 64.4%, while the degree of orientation increased firstly and then decreased. On the other hand, the lower drawing temperature (below glass transition temperature (T_g)) favored the orientation of the PA6/66 fibers. When the drawing temperature increased, the crystallinity of the fibers appeared at a maximum temperature of 40 °C.

1 Introduction

Polyamide, as a kind of semi-crystalline polymer with advantages of high strength, toughness, wear resistance and corrosion resistance, was widely applied in electronics, fibers, textiles, and other industries [1, 2]. PA6 and PA66 are two main products

W. Pan · H. Xiang (✉) · G. Zhai · M. Ni · M. Zhu (✉)
State Key Laboratory for Modification of Chemical Fibers and Polymer Materials,
College of Materials Science and Engineering, Donghua University, Shanghai 201620, China
e-mail: hengxuexiang@163.com

M. Zhu
e-mail: zhumf@dhu.edu.cn

J. Sheng · L. Kong · C. Ma
Jiangsu Haiyang Chemical Fiber Co., Ltd., Taizhou 225300, China

Y. Han (ed.), *Physics and Techniques of Ceramic and Polymeric Materials*,
Springer Proceedings in Physics 216,
https://doi.org/10.1007/978-981-13-5947-7_5

of polyamide. The strength and hardness of PA66 were better than that of PA6, while the toughness was lower. In addition, the melting point of PA66 was higher and its processing performances were restricted. Therefore, whether a new copolyamide 6/66 with combined advantages that mentioned above can be copolymerized from PA6 and PA66? In this case, the PA6/66 fibers presented higher strength and better toughness as the addition of PA66 chain segments destroyed the order of PA6 molecular chains reduced its crystallinity, lowered its melting point, and improved its processing properties. Generally speaking, PA6 has the most common crystalline forms of α and γ. The proportion of these two crystal formed in the fibers was related to the processing conditions [3].

In the α-form of PA6 crystals, the hydrogen bonds were formed between antiparallel chains which adopted a fully extended zigzag chain conformation, while in the γ-form the hydrogen bonds were formed between parallel chains [4]. In generally, the α-form, which could be obtained by annealing, slow crystallization, or treating with cyclodextrins [5, 6], was a thermodynamically stable structure with a larger Young's modulus than γ-form [7–10]. While the γ-form was always obtained via rapid melt cooling, high-speed spinning, complete blending with clay, or treating α-form PA6 with aqueous potassium iodide-iodine solution [11–13]. The two crystalline forms of α and γ in PA6 could convert into each other under certain conditions. Vasanthan [14] found that the γ/α ratio in the PA6 fibers changed from 80/20 to 25/75 under the room temperature drawing via FTIR. Two main key factors were required to change from γ-form to α-form in polyamide fibers. First, a sufficient extension of the g-phase was required to untwist the chain around the amide group. Second, an adequate molecular mobility to change the stacking in the crystallites was also needed [15]. PA66 had two crystalline forms of α and β.

In addition, it exhibited a triclinic crystal form at normal temperature and a hexagonal crystal shape at 165 °C or higher [16]. The drawing effect could not only promote the transition from γ-form to α-form, but also promote the orientation of the fibers. During the stretching process, the macromolecular chains in the amorphous region of the fibers were regularly arranged along the direction of the fiber axis under the tensile force, and the orientation was formed. As the draft ratio increased, the molecular chains were regularly aligned along the fiber axis, then the degree of orientation increased, the orientation induced crystallization where the crystallinity increased, and the breaking strength also increased which was due to the interaction of orientation and crystallization [17]. However, excessive drawing would break the molecular chains and then reduce the degree of orientation. It was generally believed that the macroscopic properties of semi-crystalline polymers were determined by microscopic features [18]. Therefore, the microstructure changes of the polymer, such as crystal structure and morphology, exhibited important guiding significance for polymer science and industry. Hence, in this paper, a simple approach was used to synthesize copolyamide 6/66, and the PA6/66 fibers were obtained by melt spinning and drawing. The effect of draft ratio and drawing temperature were also investigated and discussed by 2D-WAXD, DSC, and yarn strength tester.

2 Experimental

2.1 Material and Sample Preparation

The caprolactam (GR) and Nylon 66 salt (GR, 50 wt% in H_2O) were provided by *Pingmei shenma group nylon technology co., Ltd.* (China). Caprolactam and nylon 66 salts were added to 1 L reactor with a mass ratio of 92/8, and then the temperature raised to 170–190 °C after the reactor was filled with nitrogen. The pressure was adjusted at 0.5–2 MPa while stirring, then the temperature was raised to 230–250 °C for 1–3 h isothermy. After this, the polymerization was carried out for 2 h under reduced pressure to obtain PA6/66 resin.

The PA6/66 fibers were obtained by melt-spinning and drawing process. The spinning temperature was 280–300 °C, the draft ratio (3.0, 3.5, 4.0, and 4.5) and draw temperature (10, 20, 30, 40, and 50 °C) were used in the fiber processing.

2.2 Thermal Performance of PA6/66 Fibers

The melting crystalline enthalpy of samples was determined by differential scanning calorimetry (DSC) (Q20 DSC, USA). The fiber powder (about 5–10 mg) was tested with aluminum crucible under nitrogen atmosphere protection. The samples were heated from 0 to 260 °C at a heating rate of 10 K/min and the DSC curves were recorded. The crystallinity was calculated according to the following equation [19].

$$X_c = \frac{\Delta H}{\Delta H0} \times 100\% \qquad (1)$$

where ΔH was the melting enthalpy of the samples measured by DSC, and the ΔH_O was the melting enthalpy of the sample with the same chemical structure and 100% crystallinity. For PA6, $\Delta H_0 = 230$ J/g. As the PA66 resin had a small content of PA6/66 and its melting behavior tended to be PA6, the enthalpy of PA6 was adapted for calculation.

2.3 The Mechanical Property of PA6/66 Fibers

The mechanical property of fibers was characterized using a YG029 full-automatic single yarn strength tester. The tensile properties of PA6/66 fibers were tested in accordance with standard GB/T 14344-2008. The mechanical strength of the fibers was tested by applying a pre-tension of 0.20 ± 0.02 cN/dtex to the fibers according to standard requirements.

$$\sigma = \frac{F}{T} \tag{2}$$

where σ was the breaking strength in units of cN/dtex, F was the breaking strength of the sample, the unit was cN, and T was the average linear density measured in the same experimental sample, in units of dtex.

2.4 The Crystalline Structure of PA6/66 Fibers

The crystal structure of the PA6/66 fibers were characterized using a Xeuss 2.0 two-dimensional wide angle X-ray diffractometer (Xenocs, France) where Ga Kα radiation as the X-ray source. The X-ray wavelength (λ) was 1.34144 Å, beam center = (625.871, 929.6702), and beam size = 0.8 * 0.8 mm^2. For the 2D-WAXD measurement, the distance between sample and detector was 225.834 mm. Each 2DWAXD pattern was collected for 30 min using a detector (Pilatus3R 1M, Dectris). The interplanar spacing and crystal size of the fiber samples were calculated by Bragg's equation and Scherrer formula, respectively.

$$d_{hkl} = \frac{\lambda}{2\sin\theta_{hkl}} \tag{3}$$

$$D_{hkl} = \frac{K\lambda}{\beta_{hkl} \times \cos\theta_{hkl}} \tag{4}$$

where K was the Scherrer constant with the value of 0.89, which was related to the grain shape. θ was half of the diffraction angle; λ was the incident X-ray wavelength of 0.134144 nm; and β was the peak width at 1/2 of the diffraction peak height.

The obtained two-dimensional image was integrated by SAXS software to obtain the full width at half maximum H (F_{WHM}) of the crystallization peak. The degree of crystal orientation (Π) was calculated according to the equation.

$$\Pi = \frac{360^\circ - H}{360^\circ} \times 100\% \tag{5}$$

where Π was the degree of orientation, the fibers were fully oriented when the Π was equal to 100%, and the orientation was random when the Π was equal to 0%.

3 Results and Discussion

3.1 *Effect of Draft Ratios on the Properties of PA6/66 Fibers*

The unidirectional stretching process could increase the degree of molecules orientation and improve the mechanical properties of the polymer fibers [20, 21]. In this paper, PA6/66 fibers were treated at different draft ratios, and the effects of drawing on mechanical properties and crystal structure of the fibers were also investigated. Figure 1 showed the stress–strain curves and the mechanical performance of PA6/66 fibers at 3.0, 3.5, 4.0, and 4.5 times draft ratio. The relevant parameters such as fineness, breaking strength, and elongation at break were listed in Table 1. As seen in Fig. 1 and Table 1, with the increase of draft ratio, the breaking strength of PA6/66 fibers increased from 2.98 to 5.12 cN/dtex, while the elongation at break decreased from 41.3 to 13.8%. According to the GB/T 16603-2017, the superior product standard exhibited that the breaking strength of PA6 fibers should be higher than 4.0 cN/dtex. From the above comparison, it could be inferred that the mechanical properties of PA6/66 fibers could be easily improved by introducing PA66 into PA6. Moreover, the variation coefficient of breaking strength was less than 4.5, indicating that the fibers were relatively uniform. This was because the crystal regions

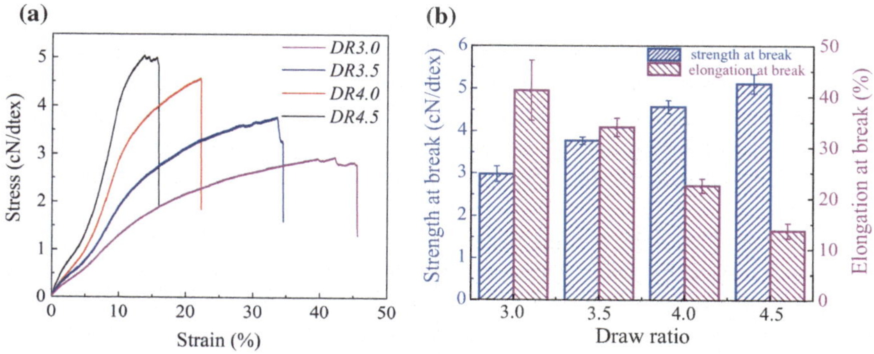

Fig. 1 **a** Stress–strain curves and **b** mechanical performance of PA6/66 fibers

Table 1 The fineness, breaking strength, and elongation at break of PA6/66 fibers under different draft ratios

Draft ratio	Fineness	Breaking strength (cN/dtex)	CV value (%)	Elongation at break (%)	CV value (%)
DR3.0	141D/33F	2.98	5.24	41.3	8.25
DR3.5	129D/33F	3.77	2.36	34.1	5.31
DR4.0	113D/33F	4.57	3.35	22.6	6.05
DR4.5	104D/33F	5.12	4.38	13.8	6.77

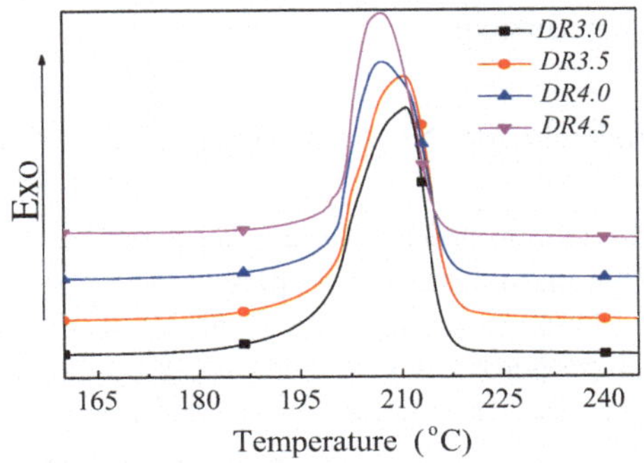

Fig. 2 Melting curves of PA6/66 fiber under different draft ratios

Table 2 Melting temperature (T_m), enthalpy (ΔH), and crystallinity (X_c) of PA6/66 fibers under different draft ratios

Draft ratios	T_m (°C)	ΔH (J/g)	X_c (%)
DR3.0	211.00	66.11	28.7
DR3.5	210.88	67.27	29.2
DR4.0	209.59	69.98	30.4
DR4.5	209.46	70.40	30.6

of the fibers were oriented by the stretching action, and the molecular chains in the amorphous region were stretched which also induced the regular alignment of the molecular chains and thereby aligned the fibers in the direction of stretching to increase the breaking strength of the fibers. Since the content of the amorphous zone was lowered, the space in the fibers that could be stretched and the elongation at break reduced.

As shown in Fig. 2, the melting behaviors of PA6/66 fibers at different draft ratios were characterized by DSC. The corresponding parameters are listed in Table 2. When the draft ratio increased from 3.0 to 4.5, the melting enthalpy of PA6/66 fibers increased from 66.11 J/g to a maximum value of 70.4 J/g, and the crystallinity increased from 28.7 to 30.6%, while the melting point decreased slightly from 211.0 to 209.5 °C. With the increase of draft ratio, the fiber regularity increased, and the sample exhibited a more symmetrical melting peak when molten. Meanwhile, the melting point tended to decrease as the destruction of the thermodynamically stable crystal structure of the fibers under a high draft ratio.

To further illustrate the effect of draft ratio on fiber crystal structure during drawing process, Fig. 3 presented the 2D-WAXD spectra of PA6/66 fibers at different draft ratios. The patterns showed the characteristic diffraction spots of (200), and (002, 202) lattice planes along the equator, and the pattern changed from a symmetrical ring to a two-point mode after drawing where the diffraction spots became brighter and more round. Figure 4 shows the integral curves of the diffracted spot along the equatorial direction obtained by integration of the spectrum in Fig. 3. At the same

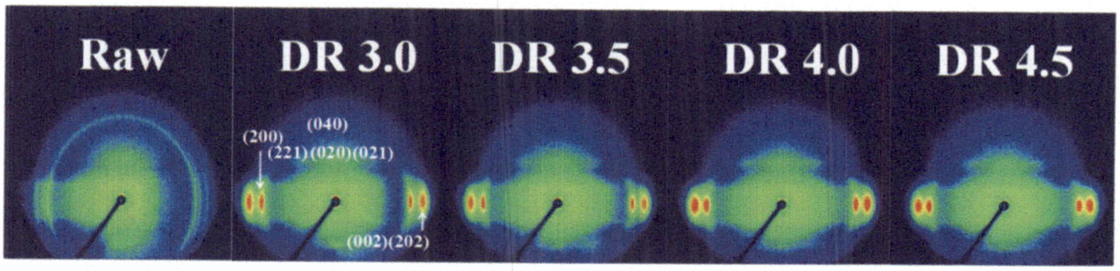

Fig. 3 2D-WAXD pattern of PA6/66 fibers under different draft ratios

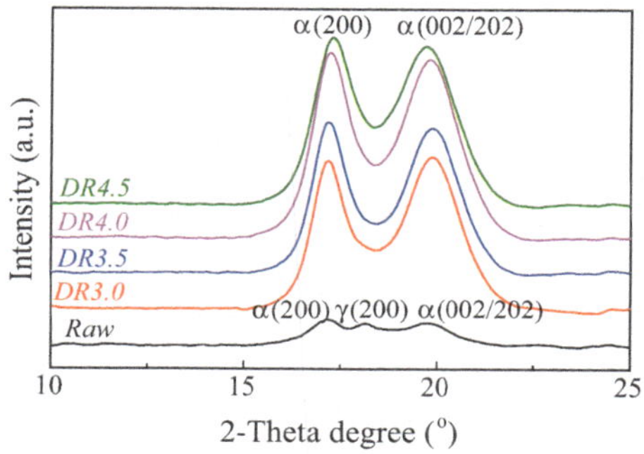

Fig. 4 1D-WAXD intensity profiles of PA6/66 fibers along the equator direction under different draft ratios

Table 3 Calculated WAXD *d* spacings, crystallite sizes, and crystallinity of PA6/66 fibers under different draft ratios

Draft ratio	(*hkl*)	2θ (°)	*d* (nm)	F_{WHM}	D_{hkl} (nm)	X_c (%)
Raw	α(200)	17.14	0.440	0.624	14.02	44.5
	γ(200)	18.15	0.417	0.515	17.02	
	α(002, 202)	19.81	0.384	0.888	9.90	
DR3.0	α(200)	17.18	0.437	0.845	10.35	56.4
	α(002, 202)	19.89	0.385	0.913	9.63	
DR3.5	α(200)	17.19	0.437	0.910	9.66	57.5
	α(002, 202)	19.85	0.383	0.971	9.06	
DR4.0	α(200)	17.26	0.436	0.929	9.42	62.3
	α(002, 202)	19.81	0.380	1.025	8.58	
DR4.5	α(200)	17.30	0.436	0.984	8.89	64.4
	α(002, 202)	19.74	0.380	1.059	8.88	

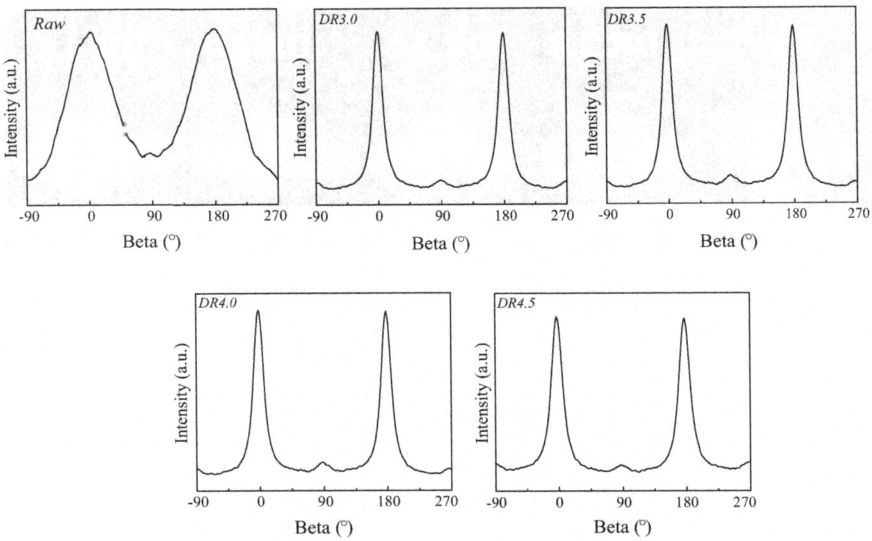

Fig. 5 WAXD orientation integral curve of PA6/66 fibers under different draft ratios

Table 4 F_{WHM} and preferred orientation based on WAXD integral curve

Draft ratio	F_{WHM} (°)	Π (%)
Raw	79.25	78.0
DR3.0	19.51	94.6
DR3.5	19.23	94.7
DR4.0	18.15	95.0
DR4.5	20.57	94.3

time, the d spacings, crystallite sizes, and crystallinity of PA6/66 fibers were also listed in Table 3. It could be observed that the diffraction peak at $2\theta \approx 18.2$ (γ(200)) disappeared after drawing, and the intensity of diffraction peak at $2\theta \approx 17.2$ (α(200)) and 19.8 (α(002, 202)) increased. This was because the extension of γ phase was sufficient to untwist the chain around the amide group under external force, then molecular mobility changed the stacking in the crystallites [15]. When the draft ratio increased, the crystallite size at α(200) decreased from 14.02 to 8.89 nm, the crystallite size at α(002, 202) decreased from 9.90 to 8.58 nm and then increased slightly, the crystallinity increased from 44.5 to 64.4%. Figure 5 shows that the WAXD orientation integral curves of PA6/66 fibers under different draft ratios, and the F_{WHM} and preferred orientation are listed in Table 4. With the increase of draft ratio, the degree of orientation increased from 78.0 to 95.0% and then decreased slightly. These results indicated that the degree of fiber crystal orientation and the crystallinity increased with the increase of the draft ratio. This result was consistent with the mechanical properties of the fibers and the DSC results. This was mainly because that the macromolecular chain in the amorphous region could be aligned to increase the crystallinity of the fibers when the draft ratio increased. On the other hand, the high-intensity crystalline region would undergo the lamellar slip under high stretching.

3.2 *Effect of Drawing Temperature on the Properties of PA6/66 Fibers*

In general, the drawing temperature of polymer fibers is above its glass transition temperature and below the softening temperature. However, the treatment temperature of the fibers prepared in this paper was lower than the glass transition temperature of PA6/66 fibers (10, 20, 30, 40, and 50 °C) due to the fact of no sufficient crystals formed inside the nascent fibers. Figure 6 exhibits the stress–strain curves and mechanical performance of PA6/66 fibers under different drawing temperatures. An interesting phenomenon that with the increase of drawing temperature from 10 to 40 °C ($<T_g$), the strength decreased slightly from 4.57 to 4.47 cN/dtex could be observed. When the drawing temperature was 50 °C ($>T_g$), the breaking strength reached the minimum value (4.26 cN/dtex). Moreover, at these drawing temperatures, the breaking strength of these fibers all exceeded 4 cN/dtex, which satisfies the requirement of superior products (Table 5).

To better explain this phenomenon, we first characterized the melting behaviors of PA6/66 fibers by DSC. Figure 7 presents the melting curves of PA6/66 fibers under different drawing temperatures. The corresponding parameters are listed in Table 6. It could be concluded that when the drawing temperature was lower than T_g, the melting enthalpy value increased from 69.98 to 71.11 J/g, and the crystallinity also increased from 30.4 to 30.9%. When the drawing temperature was higher than T_g, the enthalpy value decreased to 61.74 J/g and the crystallinity was 26.8%. This was because with

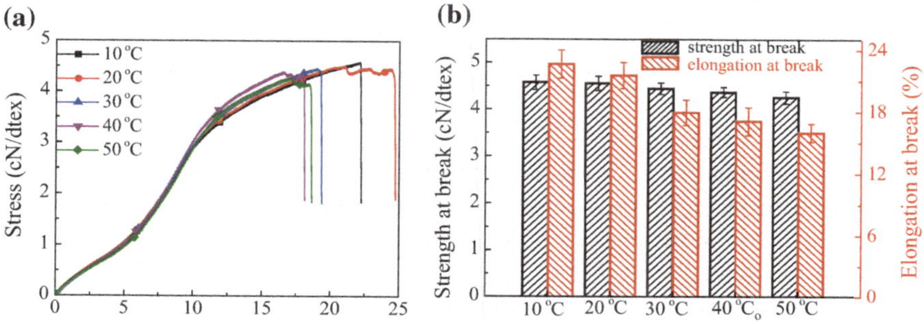

Fig. 6 Stress–strain curves (**a**) and mechanical performance (**b**) of PA6/66 fiber under different drawing temperatures

Table 5 The fineness, breaking strength, elongation at break, and CV value of PA6/66 fibers under different drawing temperatures

Drawing temperature (°C)	Fineness	Breaking strength (cN/dtex)	CV value (%)	Elongation at break (%)	CV value (%)
10	113D/33F	4.57	3.35	22.6	6.05
20	113D/33F	4.55	3.35	21.5	5.98
30	112D/33F	4.44	2.75	17.9	6.91
40	112D/33F	4.37	2.48	17.1	7.85
50	113D/33F	4.26	2.99	16.0	5.47

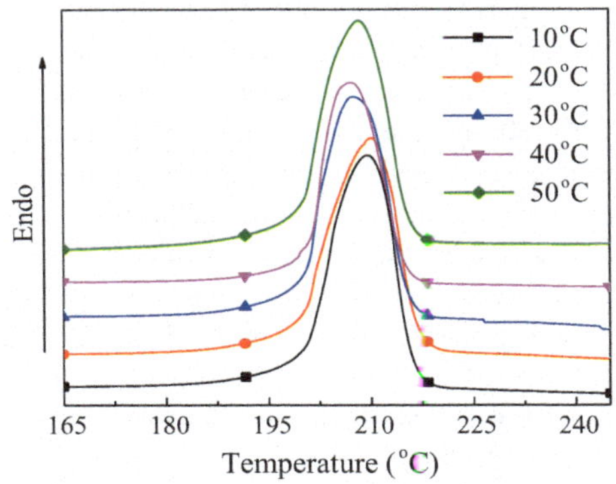

Fig. 7 Melting DSC curves of PA6/66 fibers under different drawing temperatures

Table 6 Melting temperature (T_m), enthalpy (ΔH), and crystallinity (X_c) of PA6/66 fibers under different drawing temperatures

Drawing temperature (°C)	T_m (°C)	ΔH (J/g)	X_c (%)
10	209.59	69.98	30.4
20	208.96	69.47	30.2
30	207.72	70.44	30.6
40	207.28	71.11	30.9
50	208.36	61.74	26.8

the increase of temperature, it was not beneficial to form crystal orientation for the molecular chain and crystal structure. At the same time, the increase of temperature caused an increase in the degree of molecular chain slip, reducing the mechanical properties of PA6/66 fibers.

To further illustrate the effect of drawing temperature on fiber crystal structure, Fig. 8 exhibits the 2D-WAXD spectra of PA6/66 fibers at different drawing temperatures. Figure 9 presents integral curves of the diffracted spot along the equatorial direction. The *d* spacings, crystallite sizes, and crystallinity of PA6/66 fibers are listed in Table 7. It is clearly observed that when the drawing temperature was lower

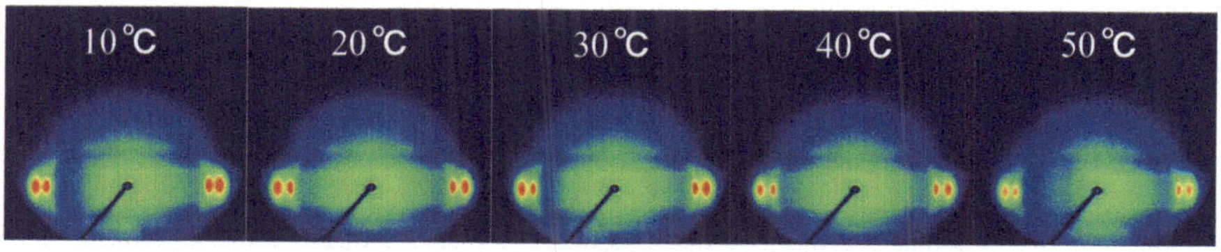

Fig. 8 2D-WAXD pattern of PA6/66 fibers under different drawing temperatures

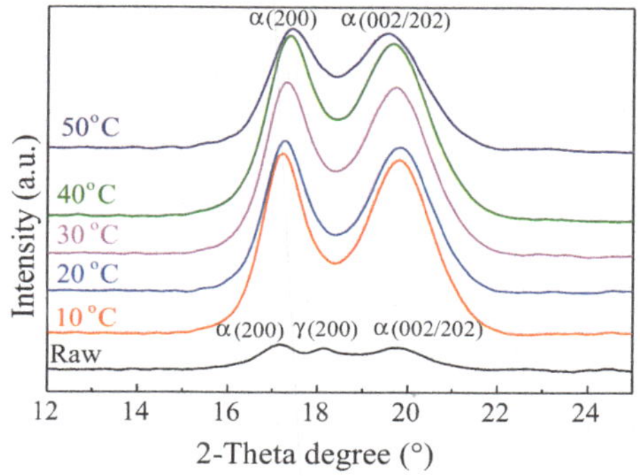

Fig. 9 1D-WAXD intensity profiles of PA6/66 fibers along the equator direction under different drawing temperatures

Table 7 Calculated WAXD *d* spacings, crystallite sizes, and crystallinity of PA6/66 fibers under different drawing temperatures

Drawing temperature (°C)	(*hkl*)	2*θ* (°)	*d* (nm)	F_{WHM}	D_{hkl} (nm)	X_c (%)
Raw	α(200)	17.14	0.440	0.624	14.02	44.5
	γ(200)	18.15	0.417	0.515	17.02	
	α(002,202)	19.81	0.384	0.888	9.90	
10	α(200)	17.26	0.436	0.929	9.42	62.3
	α(002,202)	19.81	0.380	1.025	8.58	
20	α(200)	17.21	0.434	0.879	9.96	62.5
	α(002,202)	19.83	0.380	0.904	10.38	
30	α(200)	17.33	0.432	0.855	10.24	63.7
	α(002,202)	19.72	0.379	0.897	9.81	
40	α(200)	17.38	0.431	0.911	9.61	64.1
	α(002,202)	19.65	0.382	1.018	8.64	
50	α(200)	17.45	0.428	0.921	9.51	61.2
	α(002,202)	19.54	0.384	0.999	8.80	

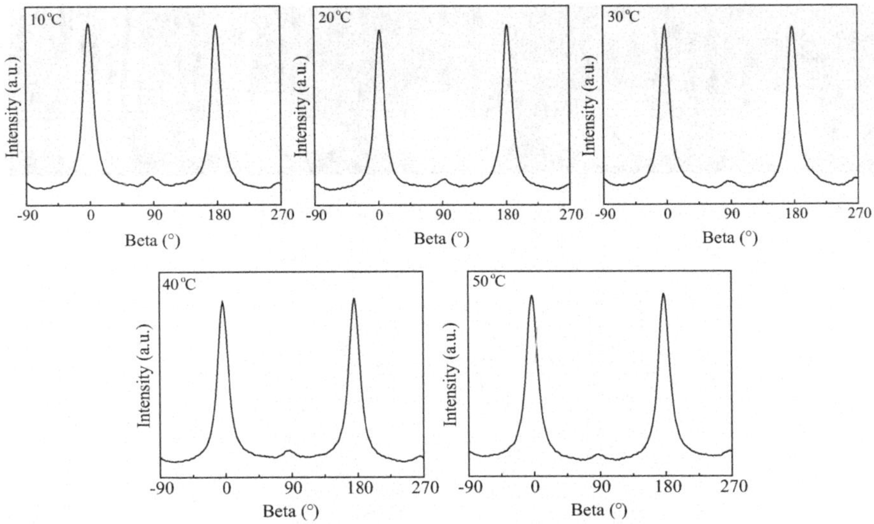

Fig. 10 WAXD orientation integral curves of PA6/66 fibers under different drawing temperatures

Table 8 F_{WHM} and preferred orientation based on WAXD integral curves

Drawing temperature (°C)	F_{WHM} (°)	Π (%)
10	18.15	95.0
20	18.50	94.9
30	18.94	94.7
40	20.31	94.4
50	16.56	92.4

than T_g, the crystallinity increased from 62.3 to 64.1% with the increase of drawing temperature. When the drawing temperature was higher than T_g, the crystallinity decreased, which was consistent with the DSC results. Figure 10 showed the WAXD orientation integral curves of PA6/66 fibers, and the F_{WHM} and preferred orientation were listed in Table 8. The degree of orientation decreased with the increase of temperature, and decreased to a minimum of 92.4% when the temperature was above T_g. Under the effect of the tensile force, the fibers were arranged along the axial direction to increase the degree of orientation and induce molecular chain crystallization as the crystallinity of fibers was little under rapid cooling conditions and a drawing at low temperatures could be achieved under a large amorphous area. It could then be illustrated that with the increase of the drawing temperature, the molecular chain segments were more likely to slip and break, resulting in a decrease in fiber orientation. At the same time, the crystallinity was improved by enhancing the movement of the molecular chain segments. Furthermore, when the drawing temperature was higher than T_g, the molecular chain moved vigorously, affecting the crystallization process of the molecule where the molecular chain tended to de-orientation process. Therefore, the macroscopic performance was that the breaking strength of the fibers decreased as the drawing temperature increased.

4 Conclusions

Copolyamide 6/66 (PA6/66) was successfully obtained and used to prepare the high-strength PA6/66 fibers by melt-spinning and drawing process. It was found that the mechanical properties of PA6/66 fibers could be easily enhanced by introducing PA66 into PA6. With the increase of draft ratio, the breaking strength of PA6/66 fibers increased from 2.98 to 5.12 cN/dtex, which was higher than the PA fiber superior product standard (GB/T 16603-2017, 4.0 cN/dtex). When the draft ratio increased, both crystallite sizes at α(200) and α(002, 202) showed clear downward trends. Compared with general polymer fibers, there was an interesting phenomenon that with the increase of drawing temperature from 10 to 40 °C ($<T_g$), the strength of PA6/66 fibers decreased slightly from 4.57 to 4.47 cN/dtex, and the degree of orientation decreased to a minimum of 92.4% when the temperature was above T_g. Despite this, the effect of the draft ratio was much higher than the drawing temperature where the mechanical strength of the PA6/66 fibers remained higher than 4.26 cN/dtex.

Acknowledgements The authors gratefully acknowledge the support from the Fundamental Research Funds for the Central Universities (2232018A3-01,2232018D3-03), the National Natural Science Foundation of China (51733002, 51603033), the Science and Technology Commission of Shanghai Municipality (16JC1400700), and the Innovation Program of Shanghai Municipal Education Commission (2017-01-07-00-03-E00055).

References

1. Z.L. Wang, J.F. Li, J. Nie, S.G. Yang, Effect of hydrothermal treatment on structure and properties of polyamide 6 fibers. Acta Polymerica Sinica **2016**(12), 1710–1716 (2016)
2. Y.F. Men, J. Rieger, Temperature dependent wide angle X-ray diffraction studies on the crystalline transition in water saturated and dry polyamide 6/66 copolymer. Eur. Polym. J. **40**(11), 2629–2635 (2004)
3. Y. Liu, L. Cui, F.X. Guan, Y. Gao, N.E. Hedin, L. Zhu, H. Fong, Crystalline morphology and polymorphic phase transitions in Electrospun Nylon-6 nanofibers. Macromolecules **40**(17), 6283–6290 (2007)
4. J.R. Xu, X.K. Ren, T. Yang, X.Q. Jiang, W.Y. Chang, S. Yang, A. Stroeks, E.Q. Chen, Revisiting the thermal transition of β-form Polyamide-6: evolution of structure and morphology in uniaxially stretched films. Macromolecules **51**(1), 137–150 (2018)
5. Y.S. Huang, F. Jiang, L.I. Jie, S.M. Peng, X. Ren, Y.H. Luo, C.W. Yi, Controllable formation of the α and γ crystalline phases of Nylon-6. J. Nat. Sci. Hunan Norm. Univ. **39**(1), 35–42 (2016)

6. M. Wei, W. Davis, B. Urban, Y.Q. Song, F.E. Porbeni, X.W. Wang, J.L. White, M.C. Balik, C.C. Rusa, J. Fox, A.E. Tonelli, Manipulation of Nylon-6 crystal structures with its α-cyclodextrin inclusion complex. Macromolecules **35**(21), 8039–8044 (2002)
7. C. Li, J.T. Yeh, Nylon 6 crystal-phase transition in nylon 6/clay/poly(vinyl alcohol) nanocomposites. J. Appl. Polym. Sci. **118**(3), 1683–1690 (2010)
8. C. Wang, S.Y. Tsou, H.S. Lin, Brill transition of nylon-6 in electrospun nanofibers. Colloid Polym. Sci. **290**(17), 1799–1809 (2012)
9. Y. Zhang, S. Liu, A. Huang, Y. Zhang, Z. Chi, J.R. Xu, Influence of thermal treatment on crystal melting behavior and crystal structure of nylon 6 and polyamide block copolymer/nylon 6 blends. Acta Polymerica Sinica **010**(2), 231–236 (2010)
10. C.F. Zhang, Y.H. Liu, S.X. Liu, H.Z. Li, K. Huang, Q.H. Pan, X.H. Hua, C.W. Hao, Q.F. Ma, C.Y. Lv, W.H. Li, Z.L. Yang, Y. Zhao, D.J. Wang, G.Q. Lai, J.X. Jiang, Y.Z. Xu, J.G. Wu, Crystalline behaviors and phase transition during the manufacture of fine denier pa6 fibers. Sci. China Chem. **52**(11), 1835 (2009)
11. M. Keizo, I. Kinzo, Effects of temperature and water on the $\gamma \rightarrow \alpha$ crystalline transition of nylon 6 caused by stretching in the chain direction. J. Polym. Sci. Part A-2 Polym. Phys. **6**(7), 1317–1329 (1968)
12. H. Haberkorn, K. Hahn, H. Breuer, H.D. Dorrer, P. Matthies, On the neck-like deformation in high-speed spun polyamides. J. Appl. Polym. Sci. **47**(9), 1551–1579 (2010)
13. C.R. And, E.B. Gowd, High-temperature X-ray diffraction studies on the crystalline transitions in the α- and γ-forms of Nylon-6. Macromolecules **34**(10), 3308–3313 (2001)
14. N. Vasanthan, D.R. Salem, FTIR spectroscopic characterization of structural changes in polyamide-6 fibers during annealing and drawing. J. Polym. Sci., Part B: Polym. Phys. **39**(5), 536–547 (2015)
15. H.L. Guo, J.Y. Wang, C.B. Zhou, W.Y. Zhang, Z.X. Wang, B.H. Xu, J.Q. Li, Y.R. Shang, J.D.C. Christiansen, D.H. Yu, Z.H. Wu, S.C. Jiang, Direct investigations of deformation and yield induced structure transitions in polyamide 6 below glass transition temperature with WAXS and SAXS. Polymer **70**(9), 109–117 (2015)
16. Z.H. Peng, Z.P. Shi, *Handbook of Plastic Industry*. Chemical Industry Press (2001)
17. Y.K. Liu, J. Liu, J. Jin, Z.C. Que, Effect of tensile heat setting on structure and properties of PET-PA/PA6 blend fibers. China Synth. Fiber Ind. **38**(3), 9–12 (2015)
18. H.Z. Li, Y.J. Wu, H. Sato, L. Kong, C.F. Zhang, K. Huang, D.L. Tao, J. Chen, X.X. Liu, Y. Zhao, Y.Z. Xu, J.G. Wu, Y. Qzaki, A new facile method for preparation of Nylon-6 with high crystallinity and special morphology. Macromolecules **42**(4), 1175–1179 (2009)
19. Z.Y. Jia, Z.Q. Cai, J.W. Chen, L. Zhang, F. Yang, X.L. Que, H.F. Bao, High efficiency toughness of aromatic sulfonamide in polyamide 6. J. Appl. Polym. Sci. **135**(30), 46527 (2018)
20. H.X. Xiang, L.L. Li, W. Chen, S.L. Yu, B. Sun, M.F. Zhu, Flame retardancy of polyamide 6 hybrid fibers: combined effects of αzirconium phosphate and ammonium sulfamate. Progr. Nat. Sci. Mat. Int. **27**(3), 369–373 (2017)
21. Z.Y. Chen, H.X. Xiang, Z.X. Hu, Z.G. Ni, M.F. Zhu, Enhanced mechanical properties of melt-spun bio-based PHBV fibers: effect of heterogeneous nucleation and drawing process. Acta Polymerica Sinica **2017**(7), 1121–1129 (2017)

Nonlinearity in Relaxor-Type Ferroelectrics Ceramics

Shen Cong, Ting-ting Qu, Cai-hong Lu, Ke Tong and Na Li

Abstract In order to research the nonlinear dielectric and piezoelectric relationships for a relaxor ferroelectrics (e.g., NBT: $Na_{0.5}Bi_{0.5}TiO_3$–$xBaTiO_3$, $x = 0.05$–0.08), structural and microstructural characterizations were identified by using X-ray diffraction (XRD) and scanning electron microscope (SEM) and P–E hysteresis loops by the ferroelectric hysteresis measurement. The results showed that the P–E loops of NBT ceramics became asymmetric and nonlinear as the field amplitudes increased. At the region of Rayleigh field, the Rayleigh coefficients α of NBT ceramics were evidently lower than conventional ceramics. The results were affected by both frequency and temperature, as the real permittivity and Rayleigh coefficient of NBT ceramics decreased with frequency increase from 0.2 to 20 Hz, and the nonlinearity of NBT ceramics became higher with the temperature increasing.

1 Introduction

The ferroelectric ceramic materials were found for the first time during a ferroelectric measurement experiment, whose container was made by the barium titanate and environment was under unnormally dielectric permanent values [1, 2]. With the excellent performance of constant for high dielectric, the ferroelectric ceramics were applied popularly in a wide range, which consist of electrical containers, high-voltage condenser, and semiconductor ceramic condenser. The composition of ferroelectric ceramics mainly involves elements lead (Pb), zirconium (Zr), barium (B), and titanium (Ti). According to the property that shows nonlinearity with electric field changing, it can be applied in dielectric amplifier [3–5].

The present research mainly focuses on the nonlinear dielectric and piezoelectric relationships for a relaxor ferroelectric (e.g., NBT: $Na_{0.5}Bi_{0.5}TiO_3$–$xBaTiO_3$, $x = 0.05$–0.08); the following objectives need to be completed one by one. Firstly, NBT ceramic samples were prepared in order to conduct structural and microstructural

S. Cong (✉) · T. Qu · C. Lu · K. Tong · N. Li
CNPC Tubular Goods Research Institute, Xi'an 710077, Shaanxi, China
e-mail: congshen@cnpc.com.cn

Y. Han (ed.), *Physics and Techniques of Ceramic and Polymeric Materials*,
Springer Proceedings in Physics 216,
https://doi.org/10.1007/978-981-13-5947-7_6

characterizations by using X-ray diffraction (XRD) and scanning electron microscope (SEM) and specific percentages of components with the technique of energy-dispersive X-ray (EDX) [6–8]. By means of ferroelectric hysteresis measurements over a range of electric field levels, frequency and temperature, to evaluate and compare results by analyzing data and drawing plots obtained from experiments [9–11]. Finally, explain observed behavior in terms of polarization mechanism and identify ferroelectric properties of NBT ceramic materials.

2 Experimental Procedure

In this part, it involves experiment procedures, some analysis techniques of experiment data, and sample preparation. The main experiment methods consist of the X-ray diffraction (XRD), the scanning electron microscopy (SEM), and the ferroelectric hysteresis measurement. In this project, most of SEM experiments were undertaken using a FEI XL-30 FEG SEM [12–15].

As we know the properties and significance of a hysteresis loop, the specific step to obtain a P–E loop by using the hardware and software and some factors during the processing of measurements will be indicated, as shown in Fig. 1.

We use the software of Lab-View to control and analyze the data from the system, because the sub-systems can be easily change when there are some mistakes happen and the operation is relatively simply to new drivers. The main operation of this system is digital storage oscilloscope (DSO), which can identify triggering and mark the level problem even without control of computer. As for the function generator, it will send the waveform amplified by the high voltage amplifier to the specimens to get the origin wave shape [16–20]. Considering time and convenience of experiments,

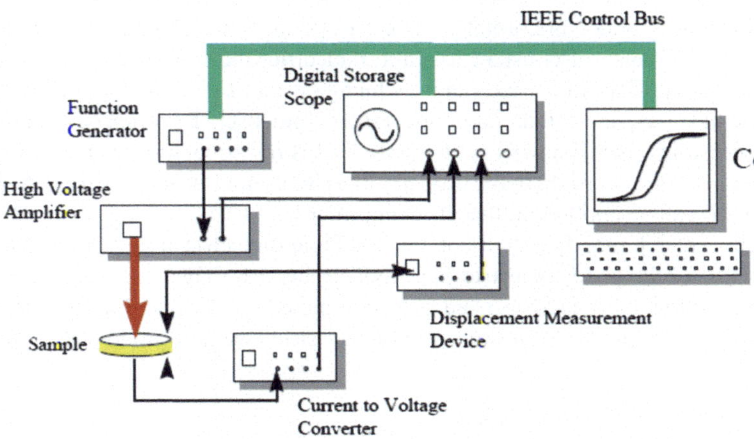

Fig. 1 Schematic illustration of P–E loop system

it is better to use computer to control the function generator, but operation by hand is also a possible way of control.

In this project, we have four NBT ($Na_{0.5}Bi_{0.5}TiO_3$-$xBaTiO_3$) specimens, x = 0.05 to 0.08 specimens separately which have varied components. The diameters of four NBT samples are, respectively, 6.599, 6.530, 6.559, and 6.585 mm. This experiment is divided into mainly three parts in terms of the effects of varied electric field amplitudes, frequency, and temperature:

Part-1 (fields): Frequency = 2 Hz; temperature = 25 °C; E0 = 0.1, 0.2, 0.3, …, 3.0 kV/mm.
Part-2 (frequency): Temperature = 25°C; frequency = 0.2, 0.5, 1, 2, 5, 10, 20 Hz; E0 = 0.1, 0.2, 0.3, …, 3.0 kV/mm.
Part-3 (temperature): Frequency = 2 Hz; temperature = 40, 60, 80, 100, 120, 140 °C; E0 = 0.1, 0.2, 0.3, …, 2.0 kV/mm.

3 Results and Discussion

3.1 X-Ray Diffraction (XRD) Results and Discussion

Figure 2 and Table 1 show the XRD pattern and reference information of poled NBT-5BT sample, and there are two reference codes which were analyzed by using 'High Score' software. To identify the crystal structure, we also need to analyze specific peaks of (111) and (200) as shown in Fig. 3.

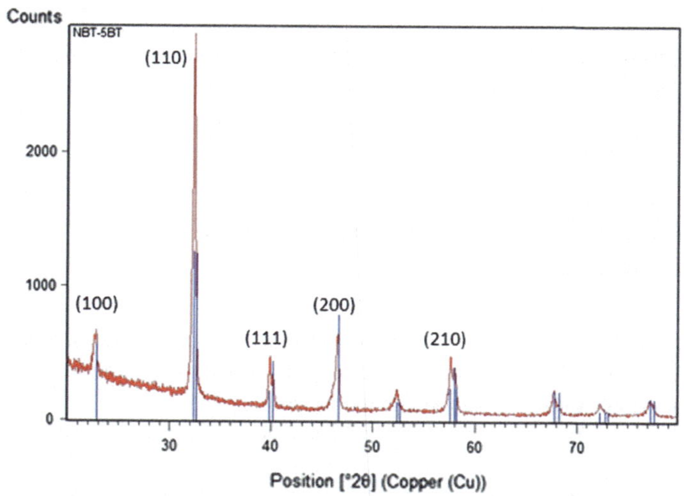

Fig. 2 XRD pattern for sample of poled NBT-5BT

Table 1 Reference information of pattern for poled NBT-5BT

Ref. code	Compound name	Chemical formula	Crystal system
04-015-0474	Sodium bismuth titanium oxide	$Na_{0.5}Bi_{0.5}TiO_3$	Rhombohedral
04-011-4032	Sodium bismuth titanium oxide	$Na_{0.5}Bi_{0.5}TiO_3$	Tetragonal

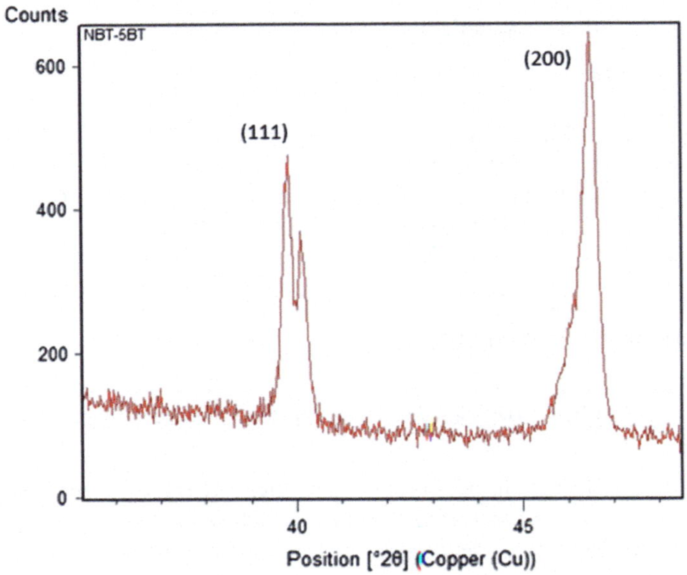

Fig. 3 (111) and (200) peaks for sample of poled NBT-5BT

By the plotting of specific peaks above, we can find out that the second peak is a single peak, and the splits in the (111) peak present the rhombohedral crystal structure.

3.2 Scanning Electron Microscopy (SEM) Results and Discussion

In this project, the technique of SEM was performed to indicate atomic number contrast by backscattered electron (BSE) mode and surface microstructure morphology by secondary electron (SE), which is shown below.

Figure 4 shows BSE images of NBT-5BT and NBT-7BT samples at magnification of 2000×, and it is really clear of grains on the microstructure surface. The grain size can be calculated in the range of 1–10 μm. As a whole, the grain morphology seems more intensive. It is evident that the grains morphology of NBT-7BT present

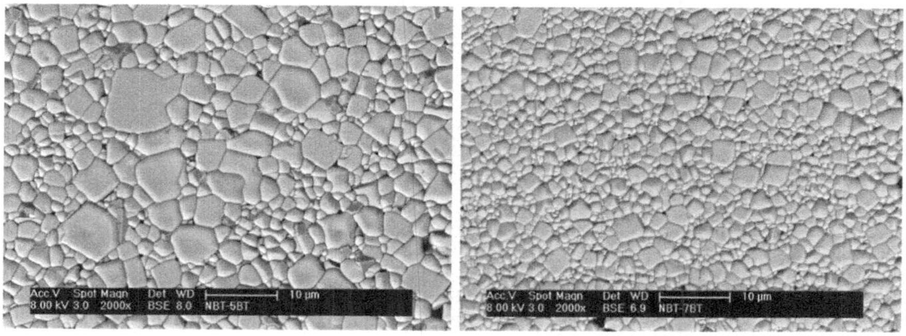

(a) NBT-5BT (b) NBT-7BT

Fig. 4 SEM images of **a** NBT-5BT and **b** NBT-7BT

more concentrated relative to NBT-5BT, and the grain size is estimated in the range of 1–3 μm. Also, there are almost no voids and shadows between grains which mean that the sample preparation before scanning was excellent.

3.3 *Ferroelectric Hysteresis Measurement Results and Discussion*

In this part of experiment, there are four NBT ($Na_{0.5}Bi_{0.5}TiO_3$–$xBaTiO_3$) specimens, $x = 0.05$–0.08. According to the previous literatures, a serious of measurements dependent on varied field, frequency and temperature, were conducted (Fig. 5).

3.4 *Field Dependence of P–E Relationships*

As we introduced above, there are four NBT ($Na_{0.5}Bi_{0.5}TiO_3$–$xBaTiO_3$) specimens, $x = 0.05$–0.08. Through a contrast of varied component NBT samples at low field amplitudes (1–2 kV/mm), they showed linear properties and the plot of NBT-5BT is thinner and indicates lower remanent and spontaneous polarizations. But at high fields (4–5 kV/mm), the plots of NBT-5BT and NBT-6BT became nonlinear and wider while 7BT and 8BT are not.

Rayleigh coefficients of varied component NBT ceramics materials are shown in Table 2, and it indicated that as the percentages of $BaTiO_3$ components increase the Rayleigh coefficient also raises up. According to related data, the Rayleigh coefficient of conventional PZT-S ceramics is normally more than 3000, while our coefficient α results of NBT samples are obviously lower than conventional ceramics.

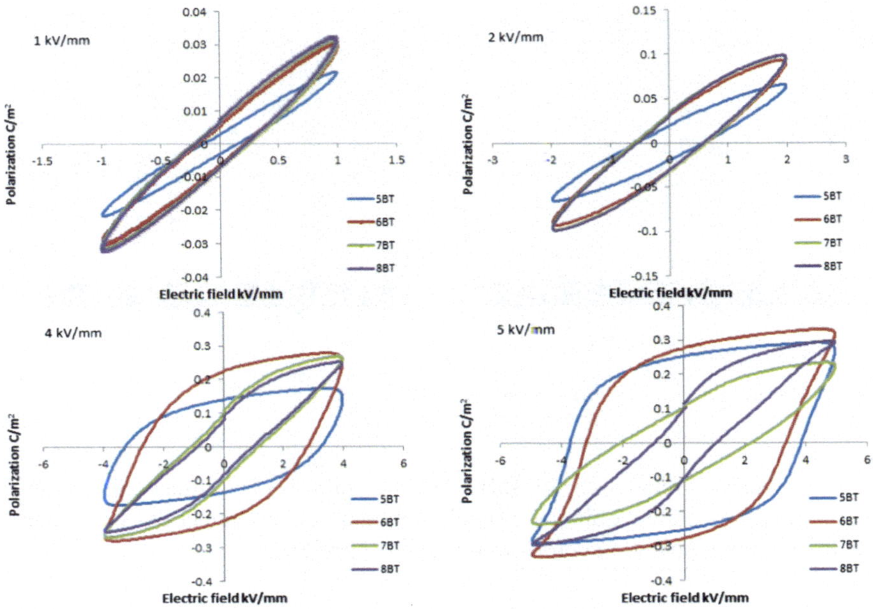

Fig. 5 P–E loops under increasing field amplitudes for NBT samples

Table 2 Rayleigh coefficients of real permittivity for NBT specimens

Specimens	α—Rayleigh coefficient (mm/kV)
NBT-5BT	1050
NBT-6BT	1750
NBT-7BT	1840
NBT-8BT	1860

This phenomenon indicated that the NBT samples have harder ferroelectricity than conventional ceramic materials.

3.5 *Frequency Dependence of P–E Relationships*

In this part, the measurements of ferroelectricity hysteresis are mainly about the influences of frequency parameter. Therefore, the other factors such as temperature and field amplitude should be kept consistent. Specially, at temperature of 25 °C field is from 0.1 to 1.0 kV/mm. Varied frequencies of 0.2, 0.5, 1, 2, 5, 10 and 20 Hz are conducted separately, and the results are also shown by relationships of permittivity-electric field and tables involves of specific Rayleigh coefficients.

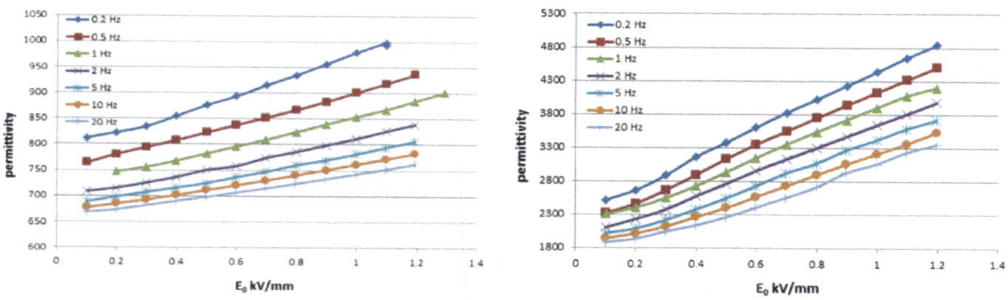

(a) NBT-5BT (b) NBT-8BT

Fig. 6 Frequency dependence of dielectric coefficients for **a** NBT-5BT and **b** NBT-8BT

Table 3 Rayleigh coefficients of real permittivity for NBT-5BT and NBT-8BT specimens at various frequencies

Frequency (Hz)	α'-NBT-5BT (mm/kV)	α'-NBT-8BT (mm/kV)
0.2	193	2170
0.5	154	2050
1	142	1840
2	122	1760
5	107	1640
10	97	1510
20	86	1420

With a combination of Fig. 6 and Table 3, the Rayleigh coefficients of NBT specimens are linearly reduced as frequency increases, and NBT-8BT has higher Rayleigh coefficients than NBT-5BT by and large. We can conclude that the higher the frequency, the higher the real permittivity and Rayleigh coefficient are. It is because that the time of domain transformation and processing of domain wall will be longer when frequency increasing.

3.6 Temperature Dependence of P–E Relationships

As like above situations, other relevant parameters like frequency are kept at 2 Hz and field at the same level. The P–E loops and relationships of $\varepsilon r'$–E0 are used to analyze influence of temperature consisting of 25, 40, 60, 80, 100, 120, and 140 °C during the measurements.

In Fig. 7, the P–E loops of NBT-8BT specimens over varied temperatures indicated that the plots of NBT become more nonlinear and symmetric with increasing temperature, which means that the nonlinearity of NBT specimens increases when the temperature gets higher.

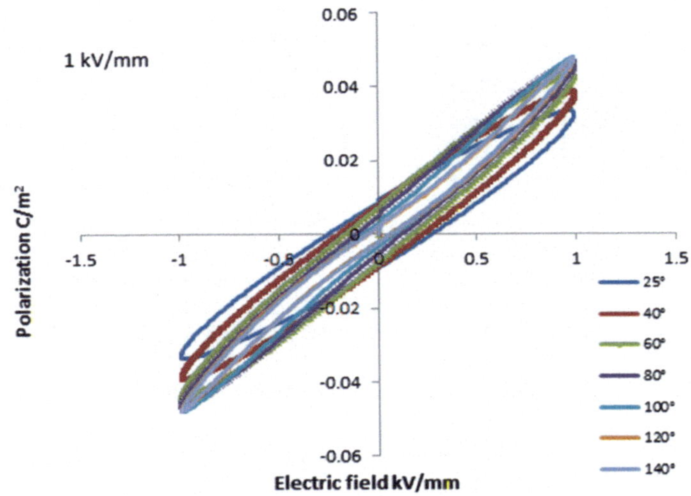

Fig. 7 P–E loops at various temperatures for NBT-8BT samples

4 Conclusions

According to a series of experiments to identify and explain dielectric behavior for relaxor ferroelectrics of NBT ceramics during ferroelectric hysteresis measurements in terms of varied field levels, frequency and temperature, we draw the following conclusions.

Within the ferroelectric hysteresis measurement, the P–E loops of NBT ceramics become asymmetric and nonlinear as the field amplitudes increase, and the coercive field, spontaneous and remanent polarizations are relatively lower. At the region of Rayleigh field, the Rayleigh coefficients α of NBT ceramics are evidently lower than conventional ceramics, which means that the ferroelectricity of NBT ceramics is harder than conventional ceramics. In terms of frequency influence, the real permittivity and Rayleigh coefficient of NBT ceramics reduce with frequency increase from 0.2 to 20 Hz. In terms of temperature effect, the nonlinearity and the Rayleigh coefficient of NBT ceramics become higher when the temperature increases.

References

1. P. Echlin, *Handbook of sample preparation for scanning electron microscopy and X-ray microanalysis* (Springer, New York, 2009)
2. G.H. Haertling, Ferroelectric ceramics: history and technology. J. Am. Ceram. Soc. **82**(4), 797–818 (1999)
3. J. Fousek, Ferroelectricity: remarks on historical aspects and present trends. Ferroelectrics **113**, 3–20 (1991)

4. L.E. Cross, R.E. Newnham, 'History of ferroelectrics', high-technology ceramics—past, present, and future. Am. Ceram. Soc. Westerville Ceram. Civilization **III**, 289–305 (1987)
5. S.B. Herner, The use of acceptor dopants to lower the loss tangent in barium strontium titanate, Doctoral dissertation, Pennsylvania State University (1993)
6. L.E. Cross, Relaxor ferroelectrics: an overview. Ferroelectrics **151**, 305–320 (1994)
7. Z.G. Ye, Relaxor ferroelectric complex perovskites: structure, properties and phase transitions. Key Eng. Mater. **155**, 81–122 (1998)
8. D. Damjanovica, A morphotropic phase boundary system based on polarization rotation and polarization extension. Appl. Phys. Lett. **97**, 062906 (2010)
9. B. Jaffe, W.R. Cook, H. Jaffe, *Piezoelectric ceramics* (Academic Press, New York, 1971)
10. J.C. Burfoot, G.W. Taylor, *Polar dielectrics and their applications* (University of California Press, Berkeley, 1979)
11. D. Damjanovic, Contributions to the piezoelectric effect in ferroelectric single crystals and ceramics. J. Am. Ceram. Soc. **88**, 2663–2676 (2005)
12. M. Stewart, M.G. Cain, D.A. Hall, *Ferroelectric hysteresis measurement and analysis* (National Physical Laboratory, Great Britain, Centre for Materials Measurement and Technology, 1999)
13. B. Lewis, Proc. Phys. Soc. **73**, 17 (1960)
14. H.J. Hagemann, Solid State Phys. J. Phys. C **11**, 3333 (1978)
15. D.A. Hall, Ferroelectrics **223**, 319 (1999)
16. S. Li, W. Cao, L.E. Cross, The extrinsic nature of nonlinear behaviour observed in lead zirconate titanate ferroelectric ceramic. J. Appl. Phys. **69**(10), 7219–7224 (1991)
17. W.G. Cady, *Piezoelectricity: an introduction to the theory and application of electromechanical phenomena in crystals* (McGraw-Hill, New York, 1946)
18. A.S. Bhalla, T.R. Oururaja, G.H. Haertling, D.M. Smyth, R.W. Vest, R.H. Tancrell, IEEE standard definitions of primary ferroelectric terms. IEEE, New York, ANSI/IEEE Std. 180-1986, pp. 1–21 (1986)
19. D.A. Hall, Nonlinearity in piezoelectric ceramics. J. Mater. Sci. **36**, 4575–4601 (2001)
20. D. Damjanovic, Hysteresis in piezoelectric and ferroelectric materials, in *Science of hysteresis*, vol. III, ed. by G. Bertotti, I. Mayergoyz (Elsevier, Armsterdam, 2005), pp. 337–465

Effect of Winding Speed on the Structure and Mechanical Properties of High-Strength Polyamide 6 Fibers

Gongxun Zhai, Hengxue Xiang, Mingda Ni, Weinan Pan, Zexu Hu and Meifang Zhu

Abstract In order to illustrate the effect of melt-spinning winding speed on the high-strength polyamide 6 fibers, the structural evolution and mechanical properties of PA6 fibers were investigated by differential scanning calorimetry (DSC), two-dimensional wide-angle X-ray diffraction (2D-WAXD), and yarn strength tester. With the increase of winding speed, the breaking strength and initial modulus of PA6 fibers increased and the elongation at breaking decreased. The fibers breaking strength reached 10.09 cN/dtex. This was due to the increase of orientation degree of PA6 fibers caused by the increasing rapid winding speed. On the other hand, the results from DSC curves of PA6 fibers indicated that the melting point of PA6 fibers maintained almost unchanged, while both the enthalpy values and the crystallinity of PA6 fibers were decreased with the increase of winding speed. Moreover, the 2D-WAXD results showed that the diffraction intensity at (200) plane and (002 + 202) plane increased with the increase of winding speed, indicating that the crystal structure transformed from γ crystal to α crystal.

1 Introduction

Polyamide 6 (PA6) fiber with high strength, good toughness, and good heat resistance [1] has become the second largest chemical fibers [2] and is widely used in the fields of textiles, packages, and other industries [3, 4]. The breaking strength of the commercial PA6 fiber is only 3.0 ~ 7.5 cN/dtex; however, the theoretical strength of the PA6 fibers can reach to 327 cN/dtex. In other words, the mechanical properties of commercial PA6 fiber are far below the ideal index [5–8]. It is generally known that there are three crystal structure types (γ, β, and α) in PA6 fiber, and the α-

G. Zhai · H. Xiang (✉) · M. Ni · W. Pan · Z. Hu · M. Zhu (✉)
State Key Laboratory for Modification of Chemical Fibers and Polymer Materials, College of Materials Science and Engineering, Donghua University, Shanghai 201620, China
e-mail: hengxuexiang@163.com

M. Zhu
e-mail: zhumf@dhu.edu.cn

Y. Han (ed.), *Physics and Techniques of Ceramic and Polymeric Materials*, Springer Proceedings in Physics 216,
https://doi.org/10.1007/978-981-13-5947-7_7

type crystallinity presents the most stable form due to fiber macromolecules that exhibit a fully extended planar zigzag conformation, and adjacent molecular chains are arranged in an antiparallel way, forming an unstrained hydrogen bond [9, 10]. Moreover, increasing draw ratio and winding speed are beneficial to the formation of α crystal in PA6 fibers [11, 12].

In general, from the point of view of high-strength polymer fibers, two methods have been mainly used to enhance the mechanical property of PA6 fibers. First, the rigid group or a size-controllable nucleating agent can be introduced into the PA6 molecular chain to eliminate the folding content of the molecular chain or to form small crystallites, thereby obtaining high-strength PA6 fibers [13]. On the other hand, the binding molecule content of the amorphous region can be increased by high stretching via reducing the molecular chain folding, and this method is also possible to increase the strength of the PA6 fibers [14, 15]. Hence, in this paper, the effects of winding speed on the mechanical properties and thermal properties of PA6 were studied, and the above results are further explained by analyzing fibers orientation and crystal structure changes.

2 Experimental

2.1 Materials

44 dtex/12f PA6 fibers were prepared with high-speed spinning (3700, 3800, 3900, 4050 m/min, respectively), and the spinning temperature was 280 ~ 300 °C.

2.2 Characterization

A YG029 auto filament SIROLAN-TENSOR was used to examine the mechanical properties of PA6 fibers, the clamping distance is 300 mm, the draw speed is 500 mm/min, the tension coefficient is 0.5, and tested 20 times for the average value. Fiber strength at breaking was calculated according to the following equation:

$$\sigma = \frac{F}{T} \tag{1}$$

where σ is strength at breaking (cN/dtex), F is breaking force (cN), and T is average line density(dtex).

The orientation of PA6 fiber was measured by an SCY-III sonic orientation instrument self-made by Donghua University. One end of a certain length PA6 fibers was clamped, and the other end was applied the tension. The average orientation F was calculated on five groups according to the equation:

$$F = 1 - (C_u/C_x)^2 \tag{2}$$

where C_u is the velocity of sonic wave in randomly oriented PA6 fiber, $C_u = 1.3$ km/s, and C_x is the speed of sonic wave in fiber.

The melting points and the enthalpy of the fibers with nitrogen as the protective gas were analyzed by TA Q20 differential scanning calorimetry (DSC, USA). 5 ~ 10 mg samples used was heated from 10 to 260 °C at a heating rate of 10 °C/min, and the crystallinity X_c is calculated by (3):

$$X_c = \frac{\Delta H_{ex}}{\Delta H_{th}} \times 100\% \tag{3}$$

where ΔH_{th} is 100% crystallinity PA6 theoretical enthalpy 230 J/g [16], and ΔH_{ex} is gained from DSC test.

The crystal structures of the fibers were characterized by RIGAKU/Max-2550 PC wide-angle X-ray diffractometer. CuKα target (voltage: 40 kV, current: 40 mA) is used in this research, spectrum wavelength 1.54056 Å, scanning angle 5° ~ 60°, scanning step 0.02°. All samples were shredded and stored in the laboratory for 2 weeks before tested.

A Franch Xenocs Xeuss 2.0 X-ray diffractometer was used to emission X-ray to scan PA6 fibers. In which Metal Jet-D2, Excillum as X-ray source, Pilatus3R 1 M, Dectris as detector, Ga/In liquid alloy as anode target, the wavelength was 0.134144 nm, the distance between the fibers sample and the detector was 225.834 mm, Beam Size = 0.8*0.8 mm^2, Beam Center = (625.871, 929.6702), Pixel size: 0.172*0.172 mm^2, Exposure time: 1800 s*4. The SAXS software was used to obtain the crystal surface integral of the obtained two-dimensional image, and then, the full width at half maximum (FWHM) H of the crystal peak can is calculated. The orientation Π of the PA6 fibers is calculated by (4):

$$\Pi = \frac{180^\circ - H}{180^\circ} \times 100\% \tag{4}$$

where Π was the degree of orientation, the fibers were fully oriented when the Π was equal to 100%, and the orientation was random when the Π was equal to 0%.

The moisture regain was examined by a method for testing chemical fibers moisture regain according to the Chinese national standard GB/T 6503-2008. The boiling water shrinkage of PA6 filament was examined according to GB/T 6505-2001 8.1, boiling water shrinkage was calculated according to (5):

$$s = \frac{L_0 - L_a}{L_0} \times 100\% \tag{5}$$

where L_0 is fibers length before heating treatment, mm, and L_a is fibers length after heating treatment, mm. The heat shrinkage of PA6 filament was examined according to Chinese national standard GB/T 6505-2001 8.2. Equation (5) is also used to calculated heat shrinkage.

3 Results and Discussion

3.1 *Effect of Winding Speed on Mechanical Properties of PA6*

The winding speed is an important factor affecting the mechanical properties of the polymer fibers [17]. Generally speaking, with the increase of the winding speed, the breaking strength increases and the elongation at breaking decreases. Figure 1 shows the stress–strain curves, breaking strength, and elongation at breaking of high-strength full drawing yarn (FDY) PA6 fibers with different winding speeds at 3700, 3800, 3900, and 4050 m/min, respectively. The corresponding parameters of PA6 fibers are listed in Table 1. As shown in Fig. 1, with the increase in winding speed, the breaking strength of fibers increased, while the elongation at breaking decreased. When the winding speed is not less than 3900 m/min, the breaking strength of the PA6 fiber reaches more than 10 cN/dtex. Moreover, it can be seen from Table 1 that with the increase of winding speeds, the initial modulus of fibers increased, from 2.47 to 3.27 GPa. These results mentioned above can be explained that the increase of winding speed directly increased the degree of crystal orientation of the fibers, improving the mechanical properties and the initial modulus of the fiber tensile curves [18].

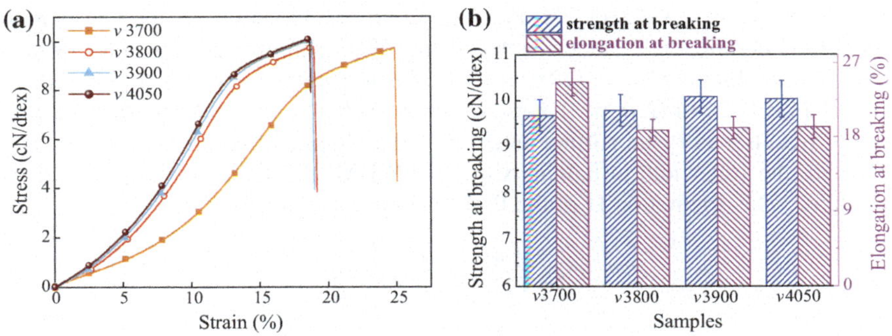

Fig. 1 Mechanical properties of high-strength PA6 FDY fibers at different winding speeds **a** stress–strain curves; **b** strength at breaking and elongation at breaking

Table 1 Mechanical properties of high-strength PA6 FDY fibers at different winding speeds

Winding speed (m/min)	Line density(dtex)	Breaking strength (cN/dtex)	CV (%)	Elongation at breaking (%)	CV (%)	Initial modulus (GPa)
3700	44.42	9.68	3.56	24.7	6.85	2.47
3800	44.34	9.79	3.54	18.8	6.99	2.76
3900	44.25	10.09	3.61	19.1	7.03	3.27
4050	44.17	10.04	4.01	19.2	7.50	3.21

Table 2 Orientation of high-strength PA6 FDY fibers at different winding speeds

Winding speed (m/min)	*C*x (km/s)	Orientation *F* (%)
3700	2.54777	73.96
3800	2.85714	79.30
3900	2.94117	80.46
4050	2.91970	80.18

Table 3 Thermal parameters of high-strength PA6 FDY fibers at different spinning speeds

Winding speed m/min	DSC heating		
	Tm (°C)	ΔH(J/g)	X_c(%)
*v*3700	220.42	81.61	33.86
*v*3800	221.11	80.30	33.32
*v*3900	221.06	75.15	31.18
*v*4050	221.14	75.09	31.16

During the drawing process, some amorphous molecular chain segments of nascent fibers will be occurred rapid molecular chain orientation along with the increase of winding speed [19]. The orientation parameters of PA6 fibers with different winding speeds are listed in Table 2. As can be seen from Table 2, the orientation F increased from 73.96 to 80.18%. This is because the orientation of crystal increases under tensile tension with the winding speed increase. The orientation of the crystal will increase the fiber breaking strength.

3.2 *Effect of Winding Speed on Thermal Properties of PA6*

Thermal properties are one of the most important properties of textile fibers, which are closely related to the wearability properties of textiles. Therefore, it is necessary to discuss the effect of winding speed on the thermal properties of PA6 fibers. Figure 2 shows the first DSC heating curves of PA6 fibers. From Fig. 2, it can be seen that the valley value of the DSC curves increased slightly along the *X* axis as the winding speed increase, indicating a slight increase in the melting point. The DSC data of PA6 fibers with different winding speeds were listed in Table 3. It can be seen that the enthalpy decreases from 81.6 to 75.1 J/g with the increase of winding speed, and the relative crystallinity calculated by (3) decreased from 33.86 to 31.16%. With the increase of fibers winding speed, the orientation of fibers increased (Table 2). During the drawing orientation process, the thermodynamic stable crystal structure is destroyed and the molecular chain lamellae appear slip-orientation, thus the enthalpy value and the crystallinity decreased [20].

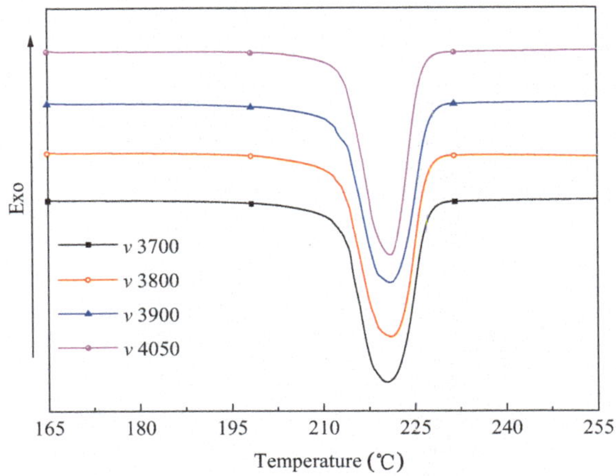

Fig. 2 DSC curves of high-strength PA6 FDY fiber at different spinning speeds

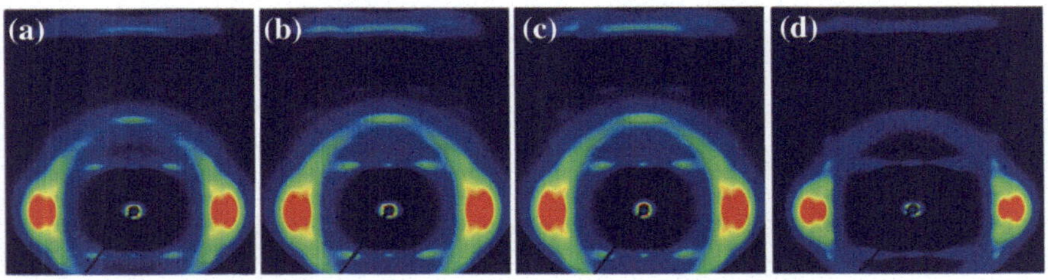

Fig. 3 WAXD spectrum of high strength PA6 FDY fiber at different spinning speeds **a** v 3700 m/min,**b** v 3800 m/min,**c** v 3900 m/min,**d** v 4050 m/min

3.3 Effect of Winding Speed on Crystalline Structures of PA6

It can be seen from the above experimental analyses that the effects of different winding speeds on the mechanical properties and thermal properties of the fibers are mainly achieved by changing the fibers structure. So the crystal structures of the fibers were characterized by WAXD and are shown in Figs. 3, 4 and Table 4. From Figs. 3 and 4, the intensity of the diffraction peak first increased and then decreased after $v = 3900$ m/min, indicating that an increase in the orientation of the fibers crystallinity as an increase in the winding speed, which is consistent with the change of orientation of the fibers structure tested by sonic velocity instrument.

As shown in Fig. 5 that as the winding speed increases, there are two peaks at both $2\theta = 17.69°$ and $\theta = 20.40°$ (α-type crystallinity), not an obvious peak at $2\theta = 18.21°$ (γ-type crystallinity) when integrated diffraction spots in WAXD spectrum with either $\beta = 0°$ or $\beta = 90°$, and the diffraction intensity of fibrous crystal planes increases at $2\theta = 17.69$ (200) and 20.40 (002 + 202), both indicated that the winding speeds can affect the crystallinity types and change from γ crystal

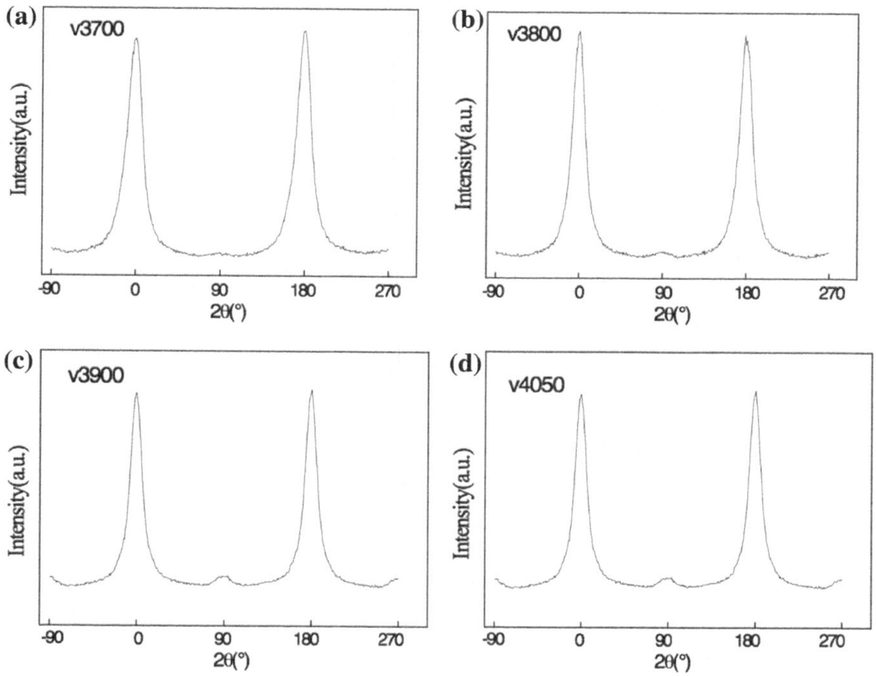

Fig. 4 2D-WAXD orientation integral curves of high-strength PA6 FDY fibers at different winding speed

Table 4 Comparison of orientation measured by two test methods of high-strength PA6 FDY fiber at different winding speed

Winding speed m/min	FWHM/°	Orientation *F* (%)	*Cx* (km/s)	Orientation *F* (%)
3700	18.19	89.9	2.54777	73.96
3800	16.34	90.9	2.85714	79.30
3900	14.88	91.7	2.94117	80.46
4050	15.11	91.6	2.91970	80.18

form to α crystal form [11]. In general, α crystal presents the most stable form due to fibers macromolecules that exhibit a fully extended planar zigzag conformation, and adjacent molecular chains are antiparallelly arranged, forming an unstrained hydrogen bond [21]. So it may be concluded that increasing winding speed can enhance the draw ratio, resulting in the form of a stable α-type crystallinity in PA6 FDY fibers.

The structural changes often cause the apparent properties of the polymer fibers; in order to discuss the application properties of high-strength PA6 fibers, the parameters including moisture regain, boiling water shrinkage, and heat shrinkage of PA6 fibers

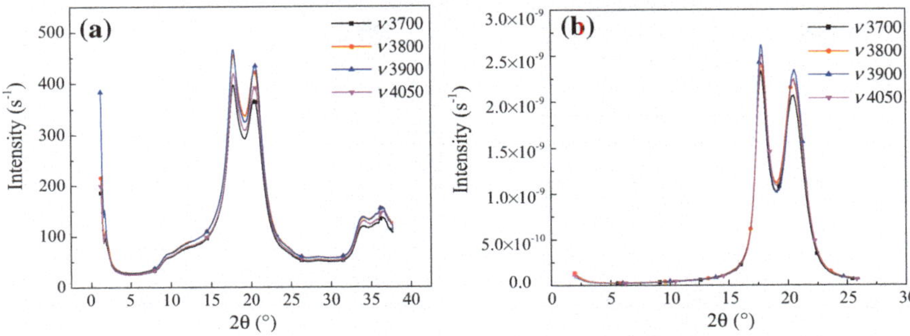

Fig. 5 Integral curves along the azimuth angle β of 2D-WAXD of high-strength PA6 FDY fiber at different winding speed **a** $\beta = 0°$ and **b** $\beta = 90°$

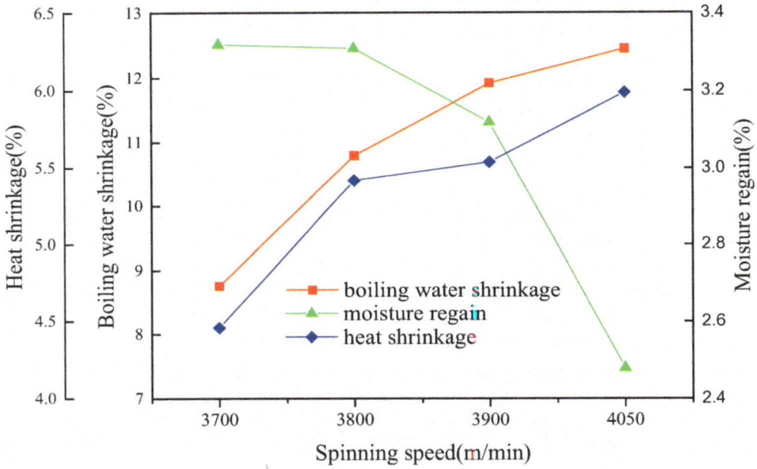

Fig. 6 Curves of moisture regain, boiling water shrinkage, and heat shrinkage of high-strength PA6 FDY fibers at different spinning speeds

are shown in Fig. 6 and Table 5. It can be seen from Fig. 6 that as the winding speed increase, both the boiling water shrinkage rate and the heat shrinkage rate increased, and the moisture regain rate decreased. According to Table 5, the boiling water shrinkage rate increased from 8.76 to 12.45%, the heat shrinkage rate increased from 4.46 to 5.99%, and the moisture regain rate decreased from 3.32 to 2.48%. This is because with the increase of winding speed, the crystallinity of the PA6 fiber changed slightly, while the degree of orientation of the fiber was markedly increased. So it can illustrate that the amorphous molecular chain and the crystal lamella were both oriented, resulting in the decreasing hydrophilicity and increasing thermal shrinkage rate of the PA6 fiber [22–24].

Table 5 Moisture regain, boiling water shrinkage, and heat shrinkage of high-strength PA6 FDY fibers at different spinning speeds

Winding speed m/min	Moisture regain (%)	Boiling water shrinkage (%)	Heat shrinkage (%)
3700	3.32	8.76	4.46
3800	3.31	10.80	5.42
3900	3.12	11.92	5.54
4050	2.48	12.45	5.99

4 Conclusions

The high-strength polyamide 6 (PA6) fibers have been successfully prepared. The effects of melt-spinning winding speed on the crystal structure and mechanical properties of PA6 fibers have been discussed. From the results, it can be concluded that with the increase of winding speed, the crystallinity of the PA6 fiber changes slightly, while the degree of orientation of the fiber markedly increases. So the breaking strength of the PA6 fiber reaches more than 10 cN/dtex, and the initial modulus of fibers increases from 2.47 to 3.27 GPa. Moreover, the amorphous molecular chains and the crystal lamella are both oriented, resulting in the decreasing hydrophilicity and increasing thermal shrinkage rate of the PA6 fiber.

Acknowledgements The authors gratefully acknowledge the support from the Fundamental Research Funds for the Central Universities (2232018A3-01, 2232018D3-03), the National Natural Science Foundation of China (51733002, 51603033), the Science and Technology Commission of Shanghai Municipality (16JC1400700), the Innovation Program of Shanghai Municipal Education Commission (2017-01-07-00-03-E00055). The authors also thank Guangdong Xinhui Meida Nylon Co., Ltd (Guangdong, China) for kindly providing the help in this study.

References

1. F.J. Li, Y. Zhou, Q.G. Huang, W.T. Yang, Synthetic and modifying technology of polyamide. Plastics **43**(4), 72–75 (2014)
2. Z.L. Wang, J.F. Li, J. Nie, S.G. Yang, Effect of hydrothermal treatment on structure and properties of polyamide 6 Fibers. Acta Polymerica Sinica **12**, 1710–1716 (2016)
3. M.I. Kohan, *Nylon Plastics Handbook*, 1st edn. (Hanser Publisher, United Kingdom, 1995)
4. G. Li, *Polymer Materials Processing Technology*, 1st edn. (China Textile & Apparel Press, Beijing, 2010)
5. V. Miri, O. Persyn, J.M. Lefebver, R. Seguela, A. Stroeks, Strain-induced disorder–order crystalline phase transition in nylon 6 and its miscible blends. Polymer **48**(17), 5080–5087 (2007)
6. J.M. Schultz, B.S. Hsiao, J.M. Samon, Structural development during the early stages of polymer melt spinning by in-situ synchrotron X-ray techniques. Polymer **41**(25), 8887–8895 (2000)
7. J.M. Samon, J.M. Schultz, B.S. Hsiao, Study of the cold drawing of nylon 6 fibers by in-situ simultaneous small- and wide-angle X-ray scattering techniques. Polymer **41**(6), 2169–2182 (2000)

8. J.M. Samon, J.M. Schultz, J. Wu, B.S. Hsiao, F. Yeh, R. Kolb, Study of the structure development during the melt spinning of nylon 6 fibers by on-line wide-angle synchrotron X-ray scattering techniques. J. Polym. Sci., Part B: Polym. Phys. **37**(12), 1277–1287 (1999)
9. Z.J. Xiong, X.N. Li, Q.X. Jia, H.H. Yin, Z.K. Yang, Progress in preparation technology of high performance polyamide 6 fibers. Chin. Polym. Bull. **4**, 30–35 (2010)
10. Y.W. Chen, R.D. Zhang, X. Chen, X.X. Feng, K. Lai, Effects of drawing on crystalline forms and mechanical properties of nylon-6 fibers. Acta polymerica sinica **12**, 1643–1647 (2014)
11. H.L. Guo, J.Y. Wang, C.B. Zhou, W.Y. Zhang, Z.X. Wang, B.H. Xu, Y.R. Li, J.D. Christiansen, D.H. Yu, Z.H. Wu, S.C. Jiang, Direct investigations of deformation and yield induced structure transitions in polyamide 6 below glass transition temperature with WAXS and SAXS. Polymer **70**(9), 109–117 (2015)
12. J.R. Xu, X.K. Ren, T. Yang, X.Q. Jiang, W.Y. Chang, S. Yang, A. Stroeks, E.Q. Chen, Revisiting the thermal transition of β-form polyamide-6: Evolution of structure and morphology in uniaxially stretched films. Macromolecules **51**(1), 137–150 (2017)
13. Y.J. Sano, Drying behavior of acetate filament in dry spinning. Drying Technol. **19**(7), 1335–1359 (2001)
14. H.P. Wang, J.M. Zhu, S.L. Hna, X.C. Hu, Theory and dynamic model on the unsymmetrical structure formation of high-speed spinning PetI. Basic theory and the influences of pump capacity and take-up speed. Synth. Fib. China **29**(5), 5–9 (2000)
15. C.W. Hao, Q.F. Ma, Z. Wang, D.T. Lu, Production process of porous super-fine nylon 6 draw textured yarn. China Synthetic Fibers Industry **34**(6), 41–43 (2011)
16. Q. Wei, D. Chionna, E. Galoppini, M. Pracella, Functionalization of LDPE by melt grafting with glycidyl methacrylate and reactive blending with polyamide-6. Macromol. Chem. Phys. **204**(8), 1123–1133 (2003)
17. K.Q. Chen, J.L. Duan, Y. Zhou, X.H. Gu, B.B. Huang, Effect of take-up velocity on structure and properties of as-spun PTT fibers. China Synth. Fib. Ind. **27**(5), 14–17 (2004)
18. X.Y. Zhang, Z.J. Pan, D.N. Qian, W.T. Shen, Configuration of electrospun polycaprolactam nano-scale fibers. J. Text. Res. **27**(7), 13–15 (2006)
19. W.Z. Zhou, J.X. Wang, L.Y. Xi, X.H. Guan, The influence of different spinning speed on the structure and properties of combined filament yarn of PET-POY. Polyester Ind. **26**(6), 12–14 (2013)
20. J. Zheng, R.W. Siegle, C.G. Toney, Polymer crystalline structure and morphology changes in nylon-6/ZnO nanocomposites. J. Polym. Sci., Part B: Polym. Phys. **41**(10), 1033–1050 (2003)
21. Z.S. Cai, *Chemistry & physics of fibers*, 1st edn. (China Textile & Apparel Press, Beijing, 2009)
22. Y.C. Li, X.H. Xu, *The Production of Polyester Filament*, 2nd edn. (China Textile & Apparel Press, Beijing, 1995)
23. H. Xiang, L. Li, W. Chen, S. Yu, B. Sun, M. Zhu, Flame retardancy of polyamide 6 hybrid fibers: Combined effects of α -zirconium phosphate and ammonium sulfamate. Prog. Nat. Sci. Mater. Int. **27**(3), 369–373 (2017)
24. Z.Y. Chen, H.X. Xiang, Z.X. Hu, Z.G. Ni, M.F. Zhu, Enhanced mechanical properties of melt-spun bio-based PHBV fibers: Effect of heterogeneous nucleation and drawing process. Acta Polymerica Sinica **7**, 1121–1129 (2017)

Natural Compounds for the Stabilization and Coloration of Polypropylene

Hui Gao, Yuxi Zhang, Weijie Xin, Zhongwei Wang, Xiaowen Huang, Xiujuan Tian, Liang Song and Qing Yu

Abstract The effects of two new natural substances (celery and Lycium chinense fruits) on the stability of polypropylene (PP) were investigated. Thermogravimetric analysis (TGA) and oxidation induction time (OIT) were conducted to investigate the thermal and thermo-oxidative stability. Melt flow rate (MFR) was tested for the processing stability. Results showed that the addition of natural substances did not change the thermal degradation behavior of PP, but greatly enhanced the thermal and thermo-oxidative stability. The temperature at 5 wt% loss ($T_{5\%}$) of PP-$L_{1.2}$ sample was 13 °C higher than pure PP. The OIT value of PP-$L_{1.2}$ sample was about 1.8 times as many as that of pure PP after the first extrusion. Moreover, the change of MFR for PP-$L_{1.2}$ sample between first and fifth extrusions was smaller than PP. As for PP-$C_{1.2}$ sample, the $T_{5\%}$ and OIT were both increased compared with PP. The change of MFR was also smaller than PP. All results confirm that celery and Lycium chinense fruits can be used as antioxidants and colorants for PP.

1 Introduction

With the continuous expansion of application field, environmental friendly polymers are required. Polyolefin is widely used in food packaging, medical equipment, containers, water pipes and toys, etc. However, susceptible degradations, caused by heat and oxygen, would result in a decline of physical and mechanical properties of polyolefin product [1–3]. In order to avoid the adverse effects, antioxidants were added to prevent thermal oxidative degradation during processing [4].

H. Gao · Y. Zhang · W. Xin · Z. Wang · X. Huang · X. Tian · L. Song · Q. Yu (✉)
Shandong University of Science and Technology, College of Materials Science and Engineering, 266590 Qingdao, China
e-mail: laura9751@163.com

Z. Wang
e-mail: wangzhongwei@fusilinchem.com

Y. Han (ed.), *Physics and Techniques of Ceramic and Polymeric Materials*, Springer Proceedings in Physics 216,
https://doi.org/10.1007/978-981-13-5947-7_8

Stabilizer packages protecting polyolefin usually contains a hindered phenolic primary antioxidants and a phosphorous or sulfuric hydroperoxide decomposer (secondary antioxidants) [5]. The main function of primary antioxidant is to react with free radicals generated during the degradation of polyolefin to prevent the oxidation of secondary antioxidants. The role of secondary antioxidants (PEPQ and 168, etc.) is to reduce hydroperoxide radicals (formed by the oxidation of polyolefin) to harmless alcohols. However, the primary antioxidants widely used in academic and industry are synthetic compounds (Irganox I1010 and 1076, etc.). These compounds have potential harm to the environment and human beings [6]. As a consequence, the attention turns to the application of natural compounds in stabilizing polyolefin. Torkelson and co-workers compared the effect of natural materials (grape pomace waste, turmeric shavings and waste, coffee grounds and orange peel waste) and Irganox I1010 in stabilizing neat low-density polyethylene (LDPE) [7]. Results have shown that in some cases, the natural materials have superior antioxidant activity compared to Irganox I1010. Földes et al. reported the antioxidative properties of β-carotene, α-tocopherol, curcumin, quercetin and dihydromyricetin in PE [8–12]. Besides these studies focused on PE, only a few ones were associated with PP in the past years [13, 14]. Taking account of the wide usage, PP was chosen as the target polymer in this study.

Celery and Lycium chinense fruits are natural edible substances, containing abundant antioxidants, such as carotene and vitamins. These compounds have benefits on human body (eliminating fatigue, protecting blood vessels, preventing arteriosclerosis, etc.) [15]. Thus, celery and Lycium chinense fruits were chosen as raw materials in this study. Solvent extraction was carried out to obtain the phenolic compounds in these two substances. The antioxidant activity of these compounds was studied in polypropylene.

2 Experimental

2.1 Materials

Polypropylene used in the experiments was PPH-T03 (melt flow rate: 4 g/10 min at 230 °C, 2.16 kg; Sinopec Beihai Refining & Chemical). Calcium stearate and PEPQ were supplied by Shanghai Aladdin Bio-Chem Technology Co., LTD. Celery and Lycium chinense fruits were purchased in the vegetable market and Tongrentang drugstore, respectively.

Fig. 1 Scheme of purification process

Table 1 Formulations and abbreviations of the samples

Samples	PEPQ (wt%)	L (wt%)	C (wt%)	CaSt (wt%)
PP	–	–	–	0.5
PP-$C_{1.2}$	0.2	–	1.2	0.5
PP-$L_{1.2}$	0.2	1.2	–	0.5

2.2 *Extractions*

Celery (C) and Lycium chinense fruits (L) were extracted with ethanol for three times at room temperature. Figure 1 shows the original powder of celery (C) and Lycium chinense fruits (L), the filtrates (C_1, C_3, C_5, L_1, L_3 and L_5) and the filter residues (C_2, C_4, C_6, L_2, L_4 and L_6). The filtrates were collected and evaporated in vacuo for further usage.

2.3 *Sample Preparations*

PP, phenol compounds, PEPQ and calcium stearate (CaSt) were mixed and extruded on a twin screw extruder. The screw temperature distribution from the hopper to the die during extrusion was 155, 165, 175, 175, 175, 175, 175, 165, 160, 160 °C, and the screw speed was 30 rpm. After cooling in water, the extrudates were granulated and subjected to another extrusion five times under the same conditions. Table 1 listed the designed formulations and the sample abbreviations.

2.4 *Sample Characterizations*

The thermal stability of the samples was tested on a Mettler Toledo TGA/DSC1 thermogravimetric analyzer. The sample (5–10 mg) was preheated to 100 °C in air or nitrogen (heating rate 10 °C/min). After 10 min, it was heated to 600 °C at a heating rate of 10 °C/min. The signal was recorded during the process. Oxidation induction time (OIT) of the sample was measured by a differential scanning calorimeter. Typical steps were as follows: (1) heating to 200 °C at a rate of 20 °C/ min in a nitrogen atmosphere (50 mL/min), (2) equilibration at 200 °C for 3 min, (3) switching to oxygen atmosphere (50 mL/min), (4) recording signal at 200 °C. The melt flow rate (MFR) was determined according to ASTM D 1238 using an XNR-400AM instrument (230 °C, 2.16 kg). The dynamic mechanical behavior of samples (3 * 6.5 * 40 mm^3) was tested on DMA1 instrument (Mettler Toledo). The tests were carried out from 0 to 100 °C at a heating rate of 5 °C/min.

3 Results and Discussion

3.1 *Thermal Stability of the Samples*

The thermal stability of pure and modified PP was investigated firstly. Figure 2 shows the TGA curves of PP, PP-$L_{1.2}$ and PP-$C_{1.2}$ samples after the first, third and fifth extrusions. The temperature at 5 wt% loss ($T_{5\%}$), maximum degradation temperature (T_{max}) and melting temperature (T_m) were listed in Table 2. Obviously, all three samples have one step degradation between 200 and 500 °C. It indicates that the thermal degradation behavior of the samples is consistent. However, the $T_{5\%}$ of PP sample is the lowest in all samples, suggesting its poorest thermal stability. T_m of PP sample is also increased after modification with the two natural substances. These results reveal that the modifications of PP with celery and Lycium chinense fruits were successful and the thermal stability of PP have been improved. Comparing the results of PP-$L_{1.2}$ and PP-$C_{1.2}$ samples, it can be concluded that Lycium chinense fruits are more effective in protecting PP against thermal degradation.

The oxidation induction time (OIT) of PP, PP-$L_{1.2}$ and PP-$C_{1.2}$ samples were measured at 200 °C. As shown in Fig. 3, the OIT values of PP-$L_{1.2}$ and PP-$C_{1.2}$ were higher than PP, indicating that the introduction of celery and Lycium chinense fruits could both improve the thermo-oxidative stability of PP matrix. Comparing the two modified samples, the OIT value of PP-$L_{1.2}$ was always higher than that of PP-$C_{1.2}$ after the first, third and fifth extrusions. It suggested that Lycium chinense fruits were more effective than celery. This result was in accordance with the TGA ones, confirming the better performance of Lycium chinense fruits in PP modification.

3.2 *Processing Stability of the Samples*

According to the TGA and DSC results, the introduction of celery and Lycium chinense fruits was benefit for the thermal and thermo-oxidative stability of polypropylene. In this section, the effects of these two natural substances on the processing stability of PP were investigated. Figure 4 illustrated the MFR of pure PP, PP-$L_{1.2}$ and PP-$C_{1.2}$ samples after the first, third and fifth extrusion. Obviously, the change

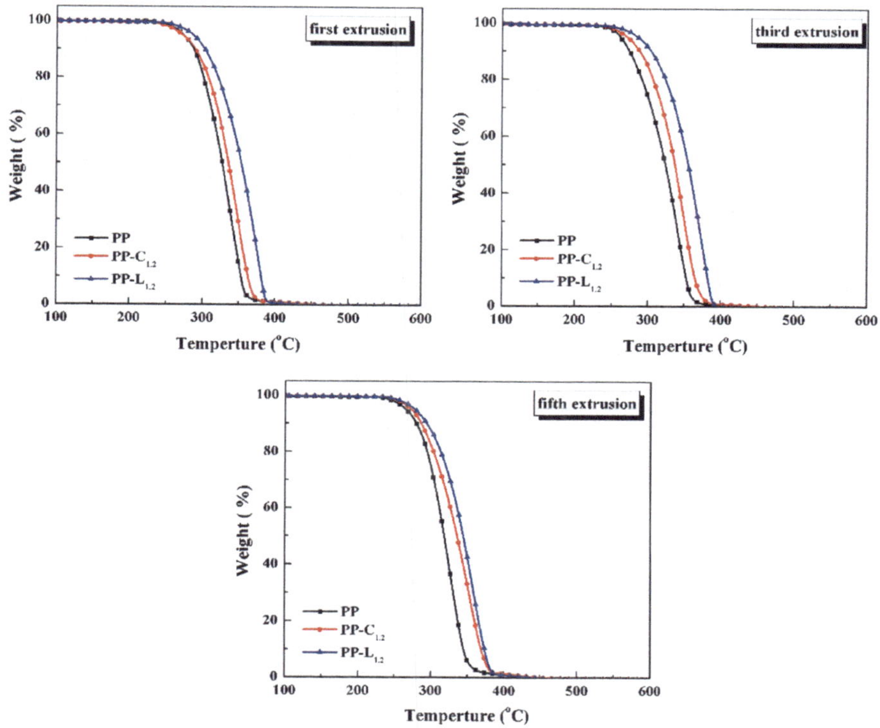

Fig. 2 Thermogravimetric curves of pure PP, PP-$L_{1.2}$ and PP-$C_{1.2}$ samples in air

Fig. 3 Oxidation induction time of pure PP, PP-$L_{1.2}$ and PP-$C_{1.2}$ samples

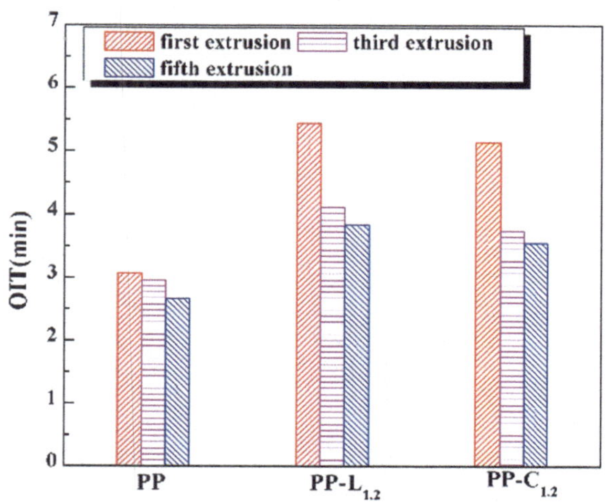

Table 2 TGA and DSC data of pure PP, $PP\text{-}L_{1.2}$ and $PP\text{-}C_{1.2}$ samples

Sample	First extrusion			Third extrusion			Fifth extrusion		
	$T_{5\%}$ (°C)	T_{max} (°C)	T_m (°C)	$T_{5\%}$ (°C)	T_{max} (°C)	T_m (°C)	$T_{5\%}$ (°C)	T_{max} (°C)	T_m (°C)
PP	269	340	169	262	342	168	261	326	167
$PP\text{-}C_{1.2}$	268	353	170	272	350	169	270	358	169
$PP\text{-}L_{1.2}$	282	377	173	274	364	172	274	353	170

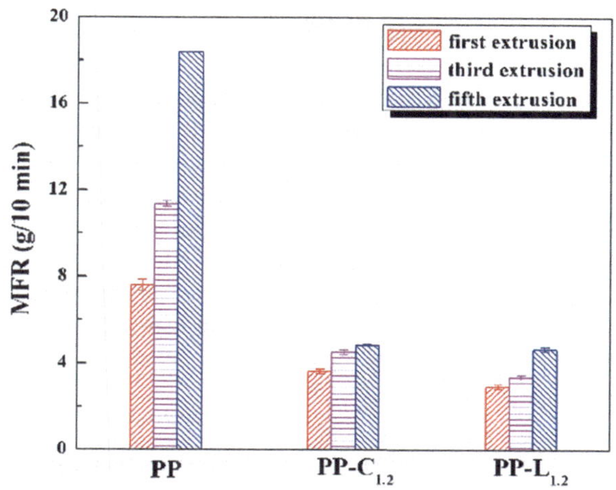

Fig. 4 MFR of pure PP, PP-$L_{1.2}$ and PP-$C_{1.2}$ samples

of MFR of PP sample was huge, indicating its terrible processing stability. The two modified samples showed much better stability. These results mean that celery and Lycium chinense fruits could be used as the processing stabilizers for PP.

3.3 Mechanical Property of the Samples

The mechanical property of pure PP, PP-$L_{1.2}$ and PP-$C_{1.2}$ samples was tested by DMA and the storage modulus (E′) values were listed in Table 3. Results showed that the storage modulus (E′) value of PP was increased after the modifications of PP with celery and Lycium chinense fruits. The enhanced stiffness of material could be attributed to the restriction of chain mobility of PP, which is caused by the formation of hydrogen bonds between additives' hydroxyl and the carbonyl groups of matrix formed by oxidation during mixing and extrusions [13].

Table 3 Storage modulus (E') of pure PP, PP-$L_{1.2}$ and PP-$C_{1.2}$ samples

Samples	E′ storage modulus (MPa) at 25 °C			E′ storage modulus (MPa) at 80 °C		
	First extrusion	Third extrusion	Fifth extrusion	First extrusion	Third extrusion	Fifth extrusion
PP	3003	882	684	2130	455	364
PP-$C_{1.2}$	3432	951	826	2453	490	425
PP-$L_{1.2}$	3453	1066	705	2451	542	368

3.4 Color of the Samples

The color of polymer product is important in some fields, while much less in other ones. Figure 5 shows the digital photos of pure PP, PP-$L_{1.2}$ and PP-$C_{1.2}$ samples. Strong coloring effect of celery and Lycium chinense fruits could be observed after the first extrusion. And the samples were much darker after the third and fifth extrusion. These results suggested that the introduction of these two new natural substances did not solve the problem of discoloration. However, from the other point of view, celery and Lycium chinense fruits could be used as the colorant for PP with enhanced thermal and processing stabilities. Such potential usages of celery and Lycium chinense fruits could be confirmed by the results of UV-Vis (Fig. 6).

Fig. 5 Digital photos of pure PP, PP-$L_{1.2}$ and PP-$C_{1.2}$ samples

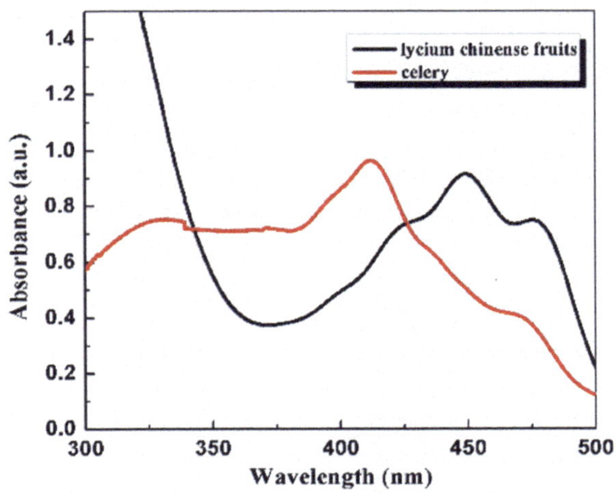

Fig. 6 Absorption spectra of celery and Lycium chinense fruits in ethanol

4 Conclusions

In this study, the extracts of celery and Lycium chinense fruits have been firstly used as antioxidants for polypropylene. These extracts are effective in enhancing the thermal, thermos-oxidative and processing stability of PP. Moreover, the two natural substances can be used as the colorant for PP.

References

1. C.C. Hsu, P.H. Geil, H. Miyaji, K. Asai, Structure and properties of polypropylene crystallized from the glassy state. Polym. Sci. Part B Polym. Phys. **24**, 2379–2401 (1986)
2. S. Costanzi, R. Farris, D. Girelli, New high performance phosphites. Polym. Degrad. Stab. **73**(3), 425–430 (2001)
3. B. Xue, K. Ogata, A. Toyota, Synthesis of polymeric antioxidants based on ring-opening metathesis polymerization (ROMP) and their antioxidant ability for preventing polypropylene (PP) from thermal oxidation degradation. Polym. Degrad. Stab. **93**, 347–352 (2008)
4. I. Kriston, O.M. Ágnes, G. Nagy, P. Staniek, E. Földes, B. Pukánszky, Melt stabilisation of Phillips type polyethylene, part I: The role of phenolic and phosphorous antioxidants. Polym. Degrad. Stab. **94**(4), 719–729 (2009)
5. B. Kirschweng, D. Tátraaljai, E. Földes, B. Pukánszky, Natural antioxidants as stabilizers for polymers. Polym. Degrad. Stab. **145**, 25–40 (2017)
6. D. Brocca, E. Arvin, H. Mosbæk, Identification of organic compounds migrating from polyethylene pipelines into drinking water. Water Res. **36**, 3675–3680 (2002)
7. K.A. Iyer, L.H. Zhang, M. Torkelson, Direct use of natural antioxidant-rich agro-wastes as thermal stabilizer for polymer: Processing and recycling. ACS Sustainable Chem. Eng. **4**, 881–889 (2016)
8. D. Tátraaljai, L. Major, E. Földes, B. Béla Pukánszky, B. Pukánszky, Study of the effect of natural antioxidants in polyethylene: Performance of β-carotene. Polym. Degrad. Stab. **102**, 33–40 (2014)
9. D. Tátraaljai, B. Kirschweng, J. Kovacs, E. Földes, B. Pukánszky, Processing stabilisation of PE with a natural antioxidant, curcumin. Eur. Polym. J. **49**(6), 1196–1203 (2013)
10. B. Kirschweng, D. Tátraaljai, E. Földes, B. Pukánszky, Efficiency of curcumin, a natural antioxidant, in the processing stabilization of PE: concentration effects. Polym. Degrad. Stab. **118**, 17–23 (2015)
11. D. Tátraaljai, E. Földes, B. Pukánszky, Efficient melt stabilization of poly-ethylene with quercetin, a flavonoid type natural antioxidant. Polym. Degrad. Stab. **102**, 41–48 (2014)
12. B. Kirschweng, K. Bencze, M. Sárközi, B. Hégely, G. Samu, J. Hári, D. Tátraaljai, E. Földes, M. Kállay, B. Pukánszky, Melt stabilization of polyethylene with dihydromyricetin, a natural antioxidant. Polym. Degrad. Stab. **133**, 192–200 (2016)
13. M. Messori, A comparative study of different winemaking by-products derived additives on oxidation stability, mechanical and thermal proprieties of polypropylene. Polym. Degrad. Stab. **149**, 9–18 (2018)
14. M.L. Xin, Y.J. Ma, W.H. Lin, K. Xu, M.C. Chen, Use of dihydromyricetin as antioxidant for polypropylene Stabilization. J. Therm. Anal. Calorim. **120**, 1741–1747 (2015)
15. R.A. Momin, M.G. Nair, Antioxidant, cyclooxygenase and topoisomerse inhibitory compounds from Apium graveolens Linn. seeds. Phytomedicine **9**(4), 312–318 (2002)

Surface Plasma Modification and Coating Properties of Quartz Fiber

Yan Xiang, Weiwei Chen, Huanwu Cheng, Aiming Bu and Yongfu Zhang

Abstract As an amorphous material, quartz fiber at high temperature is prone to cracks and protrusions on the surface. When the temperature reaches 950 °C, the crystallization behavior causes a sharp drop in mechanical properties, which greatly limits its industrial application. Aiming at this problem, this study used a liquid-phase non-electrode plasma electrolysis treatment device to achieve continuous surface treatment of quartz fiber cloth deposited alumina ceramic coating to protect its mechanical properties, at high temperatures and explored the effect of the number of processing on the morphology and properties of the coating. The test results showed that the alumina ceramic coating was successfully deposited on the surface of the insulating material quartz fiber cloth by liquid-phase plasma electrolysis technology, and the coating improved the high-temperature resistance of the quartz fiber. As the number of processing increased, the deposition amount of the coating increased, and the protective performance decreased. When the number of processing was one time, the mechanical properties of the fiber were the best, and the tensile strength after heat treatment was 54.76 MPa, which was obviously improved compared with the blank group fiber, 37.14%.

1 Introduction

Quartz fiber is a kind of high-performance inorganic fiber with excellent high-temperature insulation, ablation resistance and good chemical stability [1]. It is widely used in aerospace, high-temperature insulation and other fields [2]. However, quartz fibers are prone to crystallization at high temperatures, resulting in brittle fibers, greatly reduced mechanical properties and greatly limited the application of quartz fibers [3–5]. Studies have shown that the deposition of ceramic coating on

---

Y. Xiang · W. Chen (✉) · H. Cheng · A. Bu · Y. Zhang
Department of Materials Science and Engineering, Beijing Institute of Technology, 100081 Beijing, China
e-mail: wwchen@bit.edu.cn

Y. Han (ed.), *Physics and Techniques of Ceramic and Polymeric Materials*, Springer Proceedings in Physics 216,
https://doi.org/10.1007/978-981-13-5947-7_9

the surface of quartz fiber can improve its stability and improve the phenomenon of crystallization. It is a feasible method to solve the above problems [6].

Al_2O_3 ceramics are often used as thermal insulation coatings and superconducting insulation coatings due to their low cost, high-temperature resistance, oxidation resistance, etc. [7, 8]. Many studies have prepared Al_2O_3 ceramic coatings on the surface of quartz fibers by traditional solgel coating method to improve their high-temperature oxidation resistance and mechanical properties [9–11], but there are problems such as long cycle and high cost.

On this basis, we have found a low-cost, high-efficiency coating method for coatings with low damage to quartz fibers, namely liquid-phase plasma electrolysis technology. In this way, an alumina ceramic coating can be prepared on the surface of a 2D quartz fiber braid. The results show that the coating can improve the tensile strength of the fiber after heat treatment.

2 Experimental

2.1 Preparation of Coating

Figure 1 is a schematic diagram of the reaction of liquid-plasma electrolysis technology [12]. During the reaction, a large amount of hydrogen bubbles are present on the surface of the cathode, and the high voltage between the electrodes causes the positive ions present in the electrolyte to concentrate near the cathode, and most of them are concentrated on the surface of the bubble. This results in a high local electric field strength between the cathode and the positive charge, and as the voltage increases, the gas space within the bubble is ionized, and plasma discharge begins to occur near the cathode. In this process, along with various physical effects and chemical effects, plasma bubbles collapse and deposit on the surface of the workpiece to form a coating. The formation mechanism of amorphous oxide ceramic coating is complex and mainly involves the following reactions:

Decomposition of $AlCl_3$:

$$AlCl_3 \rightarrow Al^{3+} + Cl^- \tag{1}$$

Formation of hydroxide:

$$Al^{3+} + OH^- \rightarrow Al(OH)_3 \tag{2}$$

Dehydration of hydroxide:

$$Al(OH)_3 \xrightarrow{\text{plasma discharge}} Al_2O_3 + H_2O \tag{3}$$

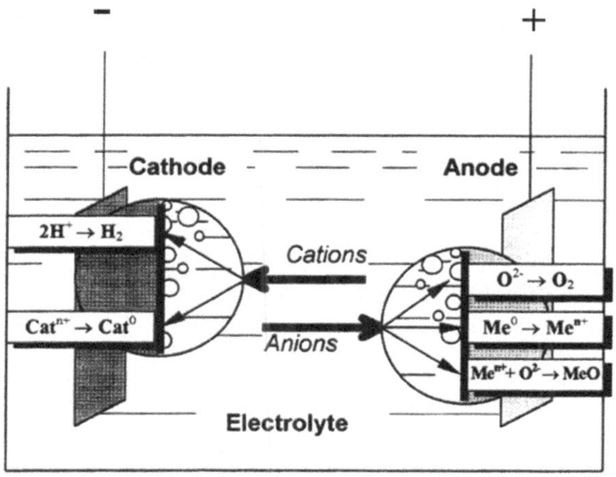

Fig. 1 Reaction schematic of plasma electrolysis. Reprinted from ref. [12], Copyright 2012, with permission from Elsevier

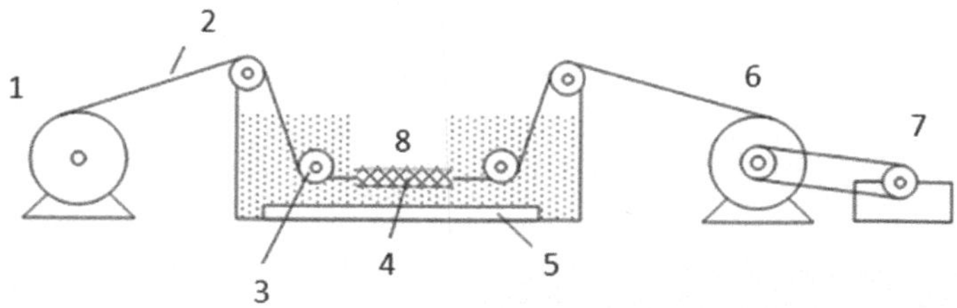

Fig. 2 The schematic diagram of a continuous non-electrode plasma electrolysis unit. 1-paying out machine; 2-quartz fiber braid; 3-pressure roller; 4-cathode; 5-anode; 6-take-up machine; 7-motor; 8-electrolyte solution. Reprinted from ref. [13], Copyright 2012, with permission from Elsevier

Figure 2 is a schematic view of the device [13], by which the continuous surface treatment of the quartz fiber can be realized and the coating deposited.

A 50 g/L $AlCl_3$ solution was added to the reaction vessel. According to the above device design, the anode and the cathode are introduced, and a voltage is applied between the two poles to generate a plasma arc region near the cathode, and the control voltage and current parameters are unchanged, and the motor is started to pass the quartz fiber at a fixed speed through the arc region, thereby realizing the coating on the fiber and uniform deposition of the surface.

2.2 *Characterization of the Coating*

The microstructure of the coating was analyzed by scanning electron microscopy. In order to determine the strength of the fiber under high-temperature conditions, the fiber was heat-treated under the heat treatment conditions of 700 °C for 30 min, then heated to 800 °C for 10 min, and then the tensile strength was tested with a stretching machine based on GB/T 7689.5-2013.

3 Results and Analysis

3.1 Energy Spectrum Analysis

Figure 3 shows the fiber element composition and morphology processed by plasma electrolysis at a constant arc voltage. It can be seen from the scanning electron micrograph that the coating deposited on the surface of the quartz fiber is distributed in a granular form due to the particle nature of the plasma. According to the energy spectrum results, the coating mainly contains Al and O elements.

Table 1 shows the energy spectrum results of Al elements under different processing times. It is found that the content of Al elements is positively correlated with the number of processing times, that is, the more the number of processing times, the more the deposition amount of the coating layer. This is due to the increase in the number of processing times, which is equivalent to prolonging the time for depositing the surface coating, and the coating growth is more sufficient.

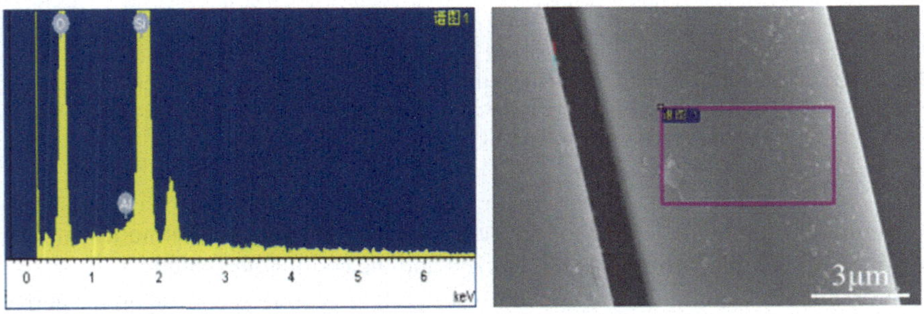

Fig. 3 Energy spectrum and morphology analysis

Table 1 Al element percentage content spectrum results

Processing times	0	1	3	5	7
Al weight %	0	0.01	0.09	0.16	0.46
Al atomic %	0	0.01	0.07	0.12	0.40

3.2 Surface Morphologies

The surface morphology of the original quartz fiber without plasma surface treatment is shown in Fig. 4a. It can be seen that the surface of the fibril is smooth and flat without obvious defects. The coating was prepared by the above experiment, and the microstructure of the coating was observed under a scanning electron microscope. Figure 4b–e is the surface shapes of the quartz fiber processed once, three times, five times, seven times, respectively.

After the surface treatment, the surface of the original smooth quartz fiber is deposited with an alumina ceramic coating, and the coating is thin and spreads along

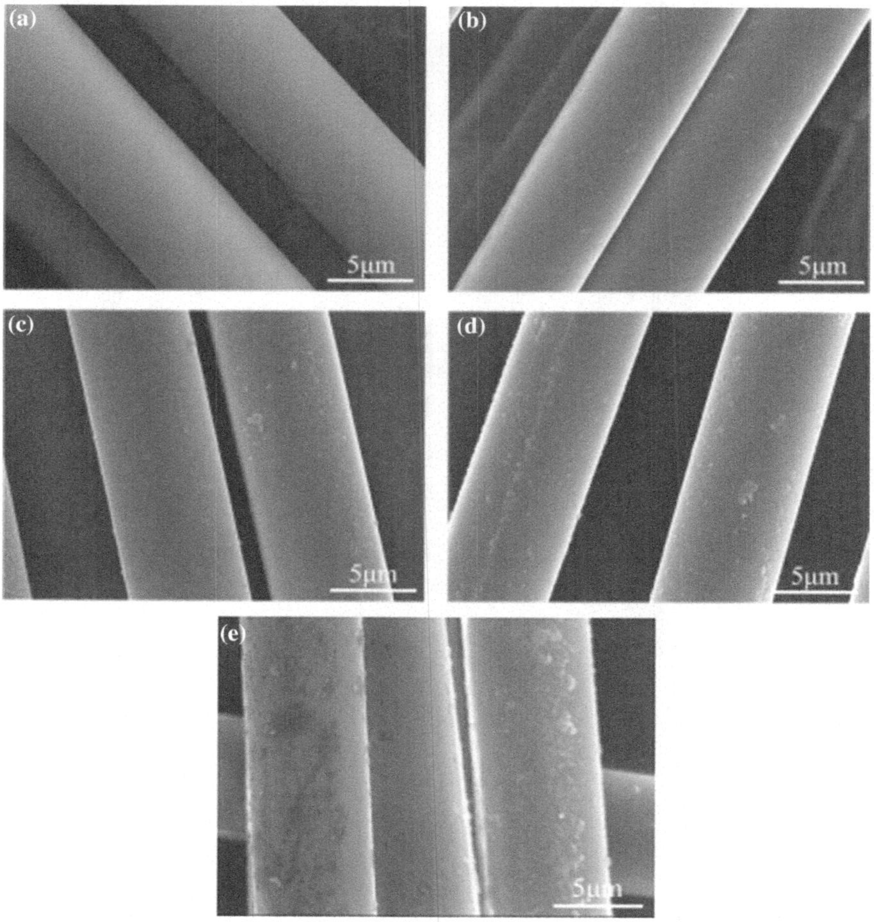

Fig. 4 SEM photographs of quartz fiber before heat treatment: **a** unprocessed; **b** processed once; **c** processed 3 times; **d** processed 5 times; **e** processed 7 times

the surface of the fiber, which is relatively uniform. With the increase of the number of processing, the coating grows more fully, and the deposition range is obviously enlarged, but there are signs of cracking and peeling at the edges. When the number of processing reaches seven times, the surface of the fiber is bombarded by the plasma arc, and holes and other defects appear.

Figure 5 shows the surface topography of a fiber sample treated with different numbers of surface treatments after heat treatment. Due to the volatilization of the surface treatment agent at a high temperature, the original quartz fiber which has lost the treatment agent has a strip-like crack, a dot or a block-like projection on the sur-

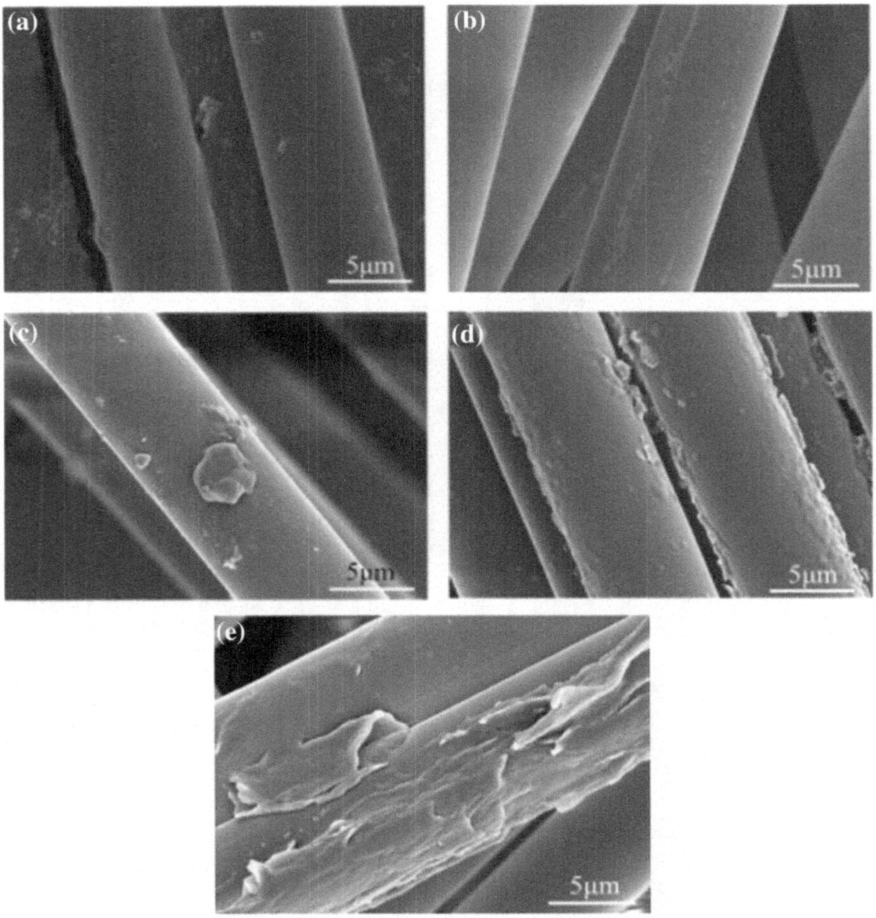

Fig. 5 SEM photographs of heat-treated quartz fiber: **a** unprocessed; **b** processed once; **c** processed 3 times; **d** processed 5 times; **e** processed 7 times

face. Under the action of thermal stress, the coating exhibits shrinkage, aggregation and flaking, which is more obvious with the increase of the number of processing.

3.3 Tensile Properties

The surface processed and unprocessed fibers are dried and placed in a SX-G12123 furnace for high-temperature treatment, and the heat treatment condition is 700 °C. After 30 min, the temperature was further raised to 800 ° C for 10 min, and it was cooled in the air to prepare a standard sample for tensile test, and the results are shown in Fig. 6 were obtained.

It can be seen from the figure that the tensile strength of the quartz fiber decreases with the increase of the number of processing after heat treatment. When the number of processing times is one, the tensile strength is 54.76 MPa, which is 37.14% higher than that of the fibrils, and the coating has the best effect on the mechanical properties of the quartz fiber after heat treatment. When the number of processing times is two or more, as the number of processing increases, the coating grows more fully, the thickness increases, and the deposition area is wider, but the uniformity of the coating is lowered. The more the number of processing times, the more severe the cracking and peeling of the coating after heat treatment. With the continuous bombardment of the plasma, each processing will cause certain damage to the matrix of the quartz fiber braid, which will reduce the mechanical properties of the fiber.

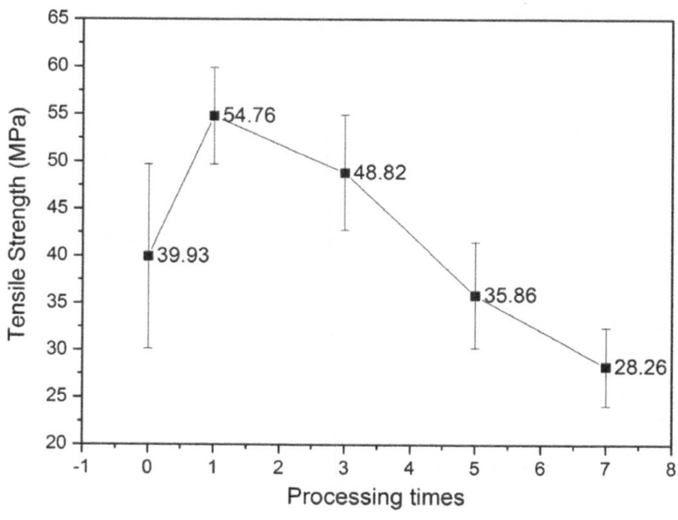

Fig. 6 Tensile strength of quartz fiber after heat treatment

4 Conclusion

In this paper, liquid-phase plasma electrolysis has been used to prepare the coating on the surface of quartz fiber braid, and the following conclusions are drawn:

(1) The alumina coating prepared on the surface of the quartz fiber braid by liquid-phase plasma electrolysis is distributed in a granular form. As the number of processing increases, the coating grows more fully.
(2) The protective effect of the coating on the mechanical properties of the quartz fiber after heat treatment decreases with the increase of the number of processing times. When the number of processing times is one, the protective effect is the most significant.

References

1. C. Chen, Characterization of new quartz fiber and preparation and properties of quartz matrix composites. National University of Defense Technology 2014
2. K. Yang, The performance of quartz fiber and its application. High temperature inorganic fiber application technology and market development seminar materials (2016)
3. Y. Zheng, Study of SiO2 nanoparticles on the improved performance of epoxy and fiber composites. J. Reinf. Plast. Compos. **24**(24), 223–233 (2005)
4. Zhen Qiang, Zhang Dahai, Wang Jinming et al., Thermal damage mechanism of quartz fiber. J. Compos. Mater. **25**(1), 105–111 (2008)
5. Xing Jianshen, Wang Shubin, Zhang Yue, Quartz fiber crystallization behavior. J. Compos. Mater. **23**(6), 75–79 (2006)
6. Wang Huaping, Liang Xiaoping, Wang Jiadan et al., Effect of acid treatment and heat treatment on the properties of alumina/quartz composite fibers. J. Tianjin Polytech. Univ. **32**(1), 10–13 (2013)
7. Chen Li, Sun Ying, Ma. Ming, Research progress of high performance fiber preforms. Progr. Mater. China **31**(10), 21–29 (2012)
8. Deng Shijun, High performance ceramic coating technology. Surface Eng. Remanuf. **3**(6), 4–5 (2003)
9. Qiu Yixia, Huo Jichuan, Lei Yonglin, Effect of phase transformation of Al_2O_3 coating on the properties of quartz fiber and its chromium phosphate composites. Funct. Mater. **42**(12), 2253–2256 (2011)
10. W. Yang, Surface treatment of quartz fiber and its composite properties. Tianjin Polytechnic University, 2012
11. Z. Wang, X. Guan, Preparation and corrosion resistance of electrophoretic deposited alumina ceramic coating. Plating and Finishing, **26**(9): 000001-4 (2007)
12. A.L. Yerokhin, X. Nie, A. Leyland, et al. Plasma electrolysis for surface engineering. Surf. Coat. Technol. **122**(2–3),73–93 (1999)
13. Y. Meng, W. Chen, H. Cheng, et al. Surface modification of quartz fiber by aqueous plasma electrolysis. Chinese Materials Conference. Springer, Singapore (2017)

Fabrication of Porous HA Ceramic Substrates by Freeze-Tape-Casting and Its Permeability Features

Mengchen Mao, Yufei Tang, Rong Xu, Kang Zhao and Xinyi Zhao

Abstract Hydroxyapatite (HA) porous substrates were fabricated by freeze-tape-casting and used as filter and biomaterials. The influence of the freezing temperature on the pore structures of HA porous substrates was investigated. The permeabilities of the samples with different solid contents and freezing temperatures were also measured. The open porosity of the fabricated HA porous substrate was high, with a closed porosity of only 5–10%, and the lamellar spaces decreased from 2.36 ± 0.53 to 1.11 ± 0.30 μm with the decreasing freezing temperature. Multilayer HA porous substrates with uniform porosity were obtained by multiple freeze-tape-castings using HA slurry with solid content of 35 wt%. The highest compressive strength of multilayer HA porous substrates was 34.98 MPa when being frozen at −45 °C. The permeability value was 4.1 μL cm^{-2} min^{-1}. The pore structure and permeability of the porous substrates can be controlled by different physical parameters and freezing conditions.

1 Introduction

Porous ceramics are widely used in catalytic reaction carriers, biomaterials and filters because of their large specific surface area, high porosity and good chemical stability [1]. At present, many preparations for porous ceramics have been reported [2]. Among these processes, freeze-casting has been more consideration, owing to its controllable porosity, pore structure and size [3]. Some scholars combine tape-casting and freeze-casting method to control the pore structure, which applies to fabricate catalyst support, solid oxide fuel cells and filters [4]. Graded pore channels and aligned acicular pores, which are very helpful for liquid filtration, can be acquired by freeze-tape-casting. However, the pore structures can be affected in many

M. Mao · Y. Tang (✉) · R. Xu · K. Zhao
Xi'an University of Technology, Xi'an 710048, China
e-mail: yftang@xaut.edu.cn

X. Zhao
The Fourth Military Medical University, Xi'an 710032, China

Y. Han (ed.), *Physics and Techniques of Ceramic and Polymeric Materials*,
Springer Proceedings in Physics 216,
https://doi.org/10.1007/978-981-13-5947-7_10

factors during this process. By changing the freezing conditions [5, 6] and physical parameters [7] (freezing temperature, freezing time and rate, solid content of slurry and so on) can influence the pore structures for porous substrates.

In addition, the type of solvent also has a great influence on the preparation of porous ceramics using freeze-casting method. Suitable solvents include water [6], camphene [7], and tert-butyl alcohol (TBA) [8]. TBA can dissolve in water infinitely and volatilize easily. When different contents of TBA and water mixed, frozen ice crystals have different morphologies. Tang et al. used the mixture of TBA-H_2O as solvent to prepare porous alumina through freeze-casting and researched the morphologies of the pore shape with the different ratio of TBA-H_2O [8]. This procedure can influence the pore structure, porosity and compressive strength of porous materials.

Hydroxyapatite (HA) is widely used in biomaterial due to its good biocompatibility [9]. Also, calcium ions, phosphate groups and a large number of hydroxyl groups existed in HA surface [10]. Owing to those special structures and compositions, HA occurs adsorption and bond with a variety of substances through electrostatic forces and hydrogen bonding [11, 12]. Therefore, HA as filter material has significant advantages and applications.

Thus, in this work, we prepared the ceramic slurry by combining the advantages of HA and TBA. Porous HA ceramic substrates with different pore structures were obtained by various freeze-tape-casting processes. The pore structure and permeability of the multilayer porous HA substrates were also characterized.

2 Experimental Procedure

2.1 *Materials and Methods*

Hydroxyapatite powder (HA, Sigma-Aldrich, USA) was used as the raw material. Tert-butyl alcohol (TBA, Tianjin Bodi Chemical Co. Ltd., China) was used as freezing medium. Dispersant polyvinylpyrrolidone (PVP, Sigma-Aldrich, USA) was used to stabilize the HA slurries. Polyvinyl butyral (PVB, Qingdao Haocheng Co., Ltd., China) was used as the binder.

HA powder was mixed with PVB (0.5 wt%) and PVP (1 wt%) in TBA solvent, then ground with a ball-mill for approximately 24 h. The prepared slurries (HA solid contents of 35, 40, 45 wt%) were poured on the copper substrates that pre-cooled at −15, −30, and −45 °C. HA slurries were cast on the substrates by tape casting with the blade clearance of 1 mm. Single-layered green bodies were formed by continued orientation after tape casting. The HA slurry with the same solid content was cast on the frozen completed single-layered green bodies by freeze-tape-casting, and this process was repeated two times to obtain uniform multilayer green bodies. The frozen bodies were dried in freeze dryer for 12 h to remove the ice and the green bodies were obtained. Samples were sintered at 1300 °C for 2 h. Single-layer porous HA ceramic substrates and multilayer porous HA ceramic substrates with uniform porosity were obtained.

2.2 *Characterization*

The morphologies of samples were characterized by the scanning electron microscopy (SEM, JSM 6390A; JEOL, Tokyo, Japan). Porosity was measured using the Archimedes principle. The lamellar spaces of porous HA substrates were determined by measuring the cross-section of the samples in the SEM images with Smile-View (Software, JEOL, Japan).The lamellar space is the distance between the pore walls. Fifty lamellar spaces were tested to obtain the average value and standard deviation. The compressive strength of the porous HA substrates was measured on a cylindrical sample with Φ10 mm using a computer servo to control the material testing machine (HT-2402-100KN, Hungta, Taiwan). Five samples were tested to obtain an average strength value. The permeability of multilayer porous HA substrate with uniform porosity was estimated by a perfusion method in vitro [13].

3 Results and Discussion

3.1 *Single-Layer Porous HA Substrates*

The solid content of the slurry has great influence on the porosity of the porous HA ceramic substrate prepared by freeze-tape-casting. The open porosity, total porosity and closed cell porosity of samples prepared with different solid content of slurries at a freezing temperature of −45 °C were measured (Table 1). With the increase in the solid content of HA slurry, the total porosity and open porosity of the HA porous substrate decreased, which are closely related to the solid content. Although directional solidification of TBA in the HA slurry leads to continuous crystallization in the freezing direction, there is still a small amount of TBA encapsulated with HA particles and crystallized into closed pores after drying. However, three groups of samples contained only 5–10% closed pores. HA particles were excluded to form HA pore walls when TBA was crystallized directionally, but HA particles are not densely packed. There is a gap between the HA particles, and some closed pores will be formed after sintering.

The cross-section morphology of single-layer porous HA ceramic substrate at different freezing temperatures was observed, as shown in Fig. 1, in which the insert images show longitudinal-section morphologies. The solid content of slurry was 40 wt%. The porous structure was a tubular straight pore channel, and its direction was parallel to the freezing direction. The pore structure was determined by the

Table 1 Porosity of single-layer porous substrates from slurries with different solid contents

Solid content (wt%)	35	40	45
Open porosity (%)	55.2222	48.6	41.6666
Closed porosity (%)	6.9	7.2	8.9
Total porosity (%)	62.1	55.8	50.5

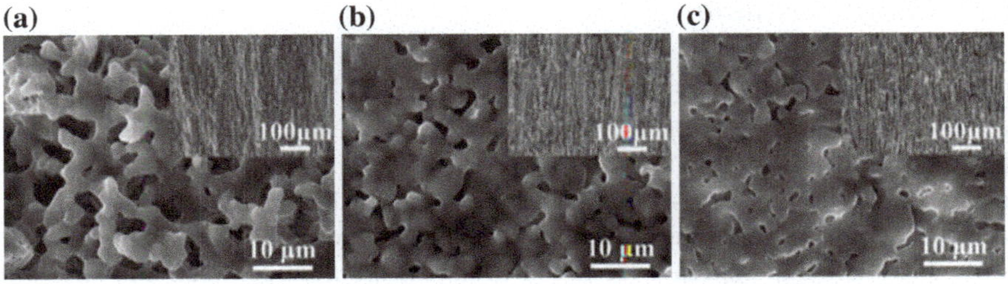

Fig. 1 Cross-section morphologies of single-layer porous HA ceramic substrates at different freezing temperatures: **a** −15 °C; **b** −30 °C; **c** −45 °C

crystal morphology of TBA and freezing temperature. The crystallization of TBA was an acicular crystal with a large aspect ratio [8]. Under directional freezing, TBA can form continuous needle-like crystals along the freezing direction and then forms needle-like straight pore channels after drying and sintering (insert images in Fig. 1). The cross-section of the TBA crystal should be a hexatomic ring shape, which was previously reported [14], but it is not obvious in Fig. 1. The reason could be the large size of the HA particles and their difficulty to be packed around the crystal in the directional growth. The selected HA powders were irregular particles, which also affected the crystalline morphology of TBA. Moreover, the lamellar spaces of the porous HA substrates decreased from 2.36 ± 0.53 to 1.11 ± 0.30 μm with the decreasing freezing temperature. The reduction in the freezing temperature leads to the increase in undercooling of HA slurry. The growth rate of TBA was fast and the crystal nucleus size was small because of the large undercooling, so the lamellar spaces were reduced.

3.2 Multilayer Porous HA Substrates with Uniform Porosity

A large thickness of a single layer by tape casting leads to the nonuniformity of green bodies due to the surface tension of HA slurries [15]. Therefore, the multilayer freeze-tape-casting technique was developed to obtain multilayer porous HA ceramic substrates. Figure 2 demonstrates the longitudinal-section morphologies of the multilayer porous HA ceramic substrates with uniform porosity from slurries with different freezing temperature when the solid content is 35 wt%. The pores formed by freeze-tape-casting, using TBA as the freezing medium, were needle-like

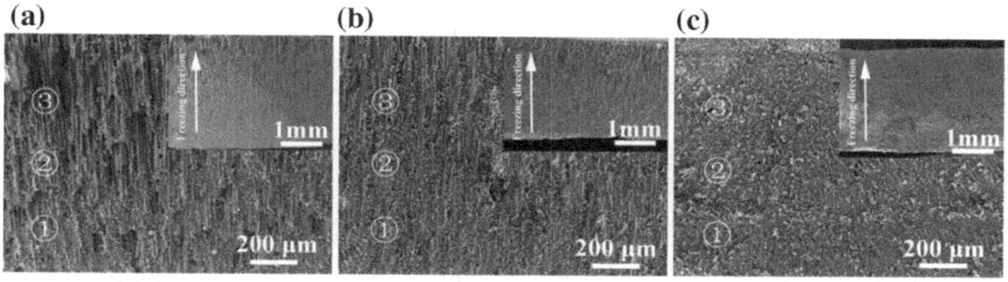

Fig. 2 Longitudinal-section morphologies of multilayer porous HA ceramic substrates with uniform porosity from slurries with different freezing temperatures when solid content of slurry is 35 wt%. **a** −15 °C, **b** −30 °C, **c** −45 °C. The number is the freeze-tape-casting order

straight pore channels and had a parallel distribution along the freezing direction. With the same solid content of slurries and freezing temperature, there was no obvious interface between the three layers of the porous HA ceramic substrates, and the porosity and lamellar spaces are basically the same. From the panorama sections (insert images of Fig. 2), the lamellar space near the cold source was small and increased along the freezing direction, as well as maintained the stability. Since there was rapid cooling of the HA slurry near the cold source, a large number of small nuclei were formed by rapid supercooling. Some heat was released during crystallization, which increased the temperature of the liquid region at the solid–liquid interface. Only part of the crystal grain continues to grow, so lamellar spaces with a small size appeared near the cold source. As the directional freezing continues, the TBA molecules overcome the resistance of the liquid–solid interface, gradually enter the crystal lattice of the TBA, and form continuous needle-like TBA crystals. Generally, the gradient structure near the cold source is inevitable, and the multiple freeze-tape-castings may result in the periodic gradient structure. However, the inhomogeneity is not found between the multilayer porous HA in Fig. 2. When the HA slurry spreads on the frozen bottom layer, the frozen part near the interface melts at the interface due to the heat transfer effect [16], resulting from the temperature of the HA slurries being at 45 °C before the second and third freeze-tape-casting. The melted part was frozen again by freezing of the cold source, and the TBA molecules in the slurry will continue to grow along the TBA crystal in the bottom layer. Finally, directional pore channels were formed, though three freeze-tape-casting cycles were carried out.

Mechanical properties of the porous ceramics were closely related to their porosity [17]. Figure 3a shows compressive strengths of multilayer HA porous substrates with uniform porosity from slurries with different solid content at different freezing temperatures. With the increase in HA content in the slurry, the compressive strength of the HA porous substrates increases because their porosity decreases. The Ryske with equation [18] shows that the compressive strength of porous materials primarily depended on the porosity. The pores reduce the cross-sectional area of the load and may induce microcracks. In addition, the size of the HA pore walls increased with the increasing solid content, which is one of the reasons to improve

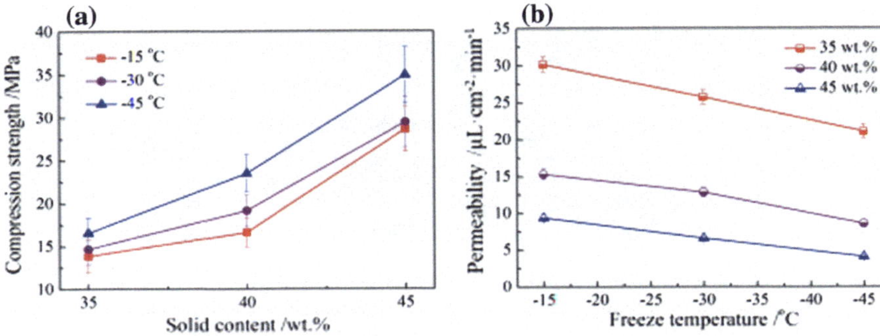

Fig. 3 **a** Compressive strengths and **b** permeability of multilayer porous HA substrates with uniform porosity from slurries with different solid contents when freezing at different temperatures

the compressive strength of porous HA. Furthermore, the compressive strength of HA porous substrates also increased with the decrease in the freezing temperature, but there was no significant difference between a freezing temperature of −15 and −30 °C. The compressive strength of HA porous substrates increased significantly, reaching 34.98 MPa, since the lamellar space from TBA crystal sublimation was small because the HA slurry was frozen at −45 °C.

The rate of water passing through the multilayer porous HA substrates was used as its permeability (Fig. 3b). With the increasing HA content of slurries, the permeability of multilayer porous HA substrates gradually decreases. The increase in HA content leads to the decrease in the porosity and induces water permeability in the unit time and unit area through the sample. The lamellar space of the porous HA substrates decreased with the freezing temperature decreasing. Water was blocked by porous substrates, and the flow rate slowed down. The permeability value of the sample reached the lowest value of 4.1 $\mu L\ cm^{-2}\ min^{-1}$. According to different applications, the permeability value of multilayer porous HA substrates could be changed by different freezing conditions and parameters, like improving the solid content of slurry or decreasing the freezing temperature and so on.

4 Summary

Porous HA ceramic substrates have been fabricated by freeze-tape-casting using TBA as freezing medium in order to obtain the acicular pore structure. The closed porosity of porous HA samples is only 5–10%, and their lamellar spaces are less than 5 μm. Multilayer porous HA ceramic substrates with uniform porosity are acquired by using the slurry of same solid content by repeatedly freeze-tape-casting. With the increase in solid content in HA slurries and decrease in freezing temperature, the compressive strength of porous HA ceramic substrates increases, but their permeabilities decrease. When the solid content is 45 wt% and the freezing temperature is −45 °C, the compressive strength of the porous HA substrate is 34.98 MPa, and its permeability value is 4.1 μL cm^{-2} min^{-1}. In the future, we expect to establish a quantitative relationship between porosity and permeability for HA substrate by controlling various types of parameters during freeze-casting.

Acknowledgements The authors would like to acknowledge the support from the National Natural Science Foundation of China (Nos. 81571014 and 51572217), China Postdoctoral Science special Foundation (2016T90937) and the Natural Science Foundation of Shaanxi Province (2016JQ5058).

References

1. W. Chen, Fabrication of porous silicon carbide ceramics with high porosity and high strength. J. Eur. Ceram. Soc. **34**(3), 837–840 (2014)
2. D. Jia, Characterization of porous silicon nitride/silicon oxynitride composite ceramics produced by sol infiltration. Mater. Chem. Phys. **124**(1), 97–101 (2010)
3. S. Deville, Freeze-casting of porous ceramics: a review of current achievements and issues. Adv. Eng. Mater. **10**(3), 155–169 (2010)
4. S.W. Sofie, Fabrication of functionally graded and aligned porosity in thin ceramic substrates with the novel freeze-tape-casting process. J. Am. Ceram. Soc. **90**(7), 2024–2031 (2010)
5. Y. Tang, Novel freeze-casting fabrication of aligned lamellar porous alumina with a centrosymmetric structure. J. Eur. Ceram. Soc. **34**(15), 4077–4082 (2014)
6. S. Deville, Ice-templated porous alumina structures. Acta Mater. **55**(6), 1965–1974 (2007)
7. G. Singh, Effect of freezing conditions on β-Tricalcium Phosphate/Camphene scaffold with micro sized particles fabricated by freeze casting. J. Mech. Behav. Biomed. Mater. **79**, 189 (2017)
8. Y. Tang, Freeze cast fabrication of porous ceramics using tert-butyl alcohol–water crystals as template. J. Eur. Ceram. Soc. **36**, 1513–1518 (2016)
9. Q. Liu, Composite biomaterials with chemical bonding between hydroxyapatite filler particles and PEG/PBT copolymer matrix. Biomaterials **19**, 1067–1072 (1998)
10. W. Ren, Bull. Chin. Ceram. Soc. **21**(1), 38–43 (2002)
11. A. Jungbauer, Biotechnol. Bioeng. **87**(3), 364–375 (2004)
12. S. Suzuki, Preparation of sintered filter for ion exchange by a doctor blade method with aqueous slurries of needlelike hydroxyapatite. Ceram. Int. **25**(3), 287–291 (1999)
13. D.H. Pashley, The effects of burnishing NaF/kaolin/glycerin paste on dentin permeability. J. Periodontol. **58**(1), 19–23 (1987)
14. L. Hu, Control of pore channel size during freeze casting of porous YSZ ceramics with unidirectionally aligned channels using different freezing temperatures. J. Eur. Ceram. Soc. **30**(16), 89–3396 (2010)
15. Q. Dong, An, optimization of the tape casting slurries for high-quality zirconia substrates. Ceram. Int. **43**, 16943–16949 (2017)
16. Y. Tang, Two-step freeze casting fabrication of hydroxyapatite porous scaffolds with bionic bone graded structure. Ceram. Int. **39**(8), 703–779 (2013)
17. J.C. Le Huec, Influence of porosity on the mechanical resistance of hydroxyapatite ceramics under compressive stress. Biomaterials **16**, 113–118 (1995)
18. R.W. Rice, Comparison of stress concentration versus minimum solid area based mechanical property-porosity relations. J. Mater. Sci. **28**, 2187–2190 (1993)

Microstructure Analysis of SiC Ceramics and SiC_f/SiC Composites by Diffusion Bonding

Xin-Lei Jia, Jian-Cheng Liu, Ying Meng, Shi-Dong Kang and Xin-Hao Jiang

Abstract With the excellent high temperature performance and irradiation stability, the SiC_f/SiC composites are considered as a kind of promising new fuel cladding material. The joint between the end plug and cladding is one of the key problems affecting its application. In this research, the connection between SiC_f/SiC composites was realized by diffusion bonding, and the microstructure and reaction products in the joint area were analyzed by SEM and XRD. The results showed that continuous dense interface layer could be formed at the junction by the chemical reaction between Ti and silicon carbide, and the interface reaction products were TiC and Ti_5Si_3. The interface layer formed a sawtooth structure at the junction with SiC_f/SiC composites. The chemical reactivity of the SiC fiber was weaker than that of the substrate, and there were gaps in the interface layer, which limited the diffusion of the interface reaction to the composite materials. There were long and thin interfacial phases in the pores of the materials, which formed strong mechanical occlusion with the SiC_f/SiC composites and improved the degree of interfacial layer bonding.

1 Introduction

Because of the characteristics such as high strength, low density, high temperature resistance and low chemical activity, SiC_f/SiC composite is a very promising high temperature structure material, which combines the advantages of high performance SiC fiber and SiC matrix. With a wide application prospect in the fields of national defense, aerospace, energy and other fields, SiC_f/SiC composites can be used as the rocket combustion chamber lining, aero turbine blades, nuclear fuel cladding, nuclear reactor vessel wall and other high temperature components [1–3]. However, SiC is a compound based on covalent bonds, it is difficult to produce large size and complex parts because of its inherent brittleness which leads to poor processing performance. Therefore, the joining between SiC_f/SiC composites and the joining

X.-L. Jia (✉) · J.-C. Liu · Y. Meng · S.-D. Kang · X.-H. Jiang
China North Nuclear Fuel Co. Ltd., Baotou 014035, China
e-mail: jiaxinlei2009@163.com

Y. Han (ed.), *Physics and Techniques of Ceramic and Polymeric Materials*,
Springer Proceedings in Physics 216,
https://doi.org/10.1007/978-981-13-5947-7_11

between SiC_f/SiC composites and heterogeneous materials (such as metals) have become the key to the wide application of SiC_f/SiC composite materials. For the successful development of SiC_f/SiC fuel cladding, joining of the cladding to the end plug is recognized as a technological hurdle. At present, the joining technology [4, 5] of SiC composites mainly uses the connection method between SiC sintered ceramics and metal or SiC sintered ceramics themselves.

There are various options for joining SiC to SiC [6], such as metallic braze-based joining [7, 8], glass ceramics joining [9, 10], solid-state diffusion bonding [11], SiC pre-ceramics precursors [12] and other methods [13–17]. Brazing with various metallic filler materials has been widely studied for joining of SiC. Brazing typically enables a relatively low processing temperature and low pressure or pressureless joining. Since the filler material melts or softens during heating, it is relatively easy to apply the filler to components with complex geometries. Because of different thermal expansion coefficient from brazed joint and parent material, the joint will be damaged by stress concentration in the contact area. Glass ceramics bonding is demonstrated without applying external stress at relatively low temperatures (~1500 °C). The eutectic liquid or softened glass spreads across the joint plane, which enables good contact between the tube and end-cap without applied stress during joining. Application of this method in nuclear materials will be affected by the presence of sintering aids. Joining using SiC pre-ceramics precursors provides an SiC-based joint layer at low processing temperatures and very low pressures (~0.01 MPa [18]). A limitation of this method is the difficulty to obtain a dense joint layer. Solid-state diffusion bonding with an interlayer typically uses a refractory metallic foil such as titanium and molybdenum. The advantage of this joining method is the relatively low joining temperature (~1500 °C) and low pressure, which makes this joining method attractive for the SiC_f/SiC fuel cladding.

2 Experimental

2.1 Materials

The sintered SiC ceramics used in this study are prepared in our group, which are sintered at 1900 °C for 60 min in Ar atmosphere using a little amount of Y_2O_3 and Al_2O_3 as the sintering additives. The considered material is a SiC_f/SiC composite tube with 12 mm average diameter based on SiC fibers (Amosic-SiC, Xiamen University), 2D tube preforms are coated with a 200 nm pyrocarbon interphase and the fiber preform is densified by the SiC matrix, using for both the chemical vapor infiltration (CVI) process. The titanium (impurity < 0.01%) sheet is processed into a thin slice of $\Phi 12 \times 0.5$ mm.

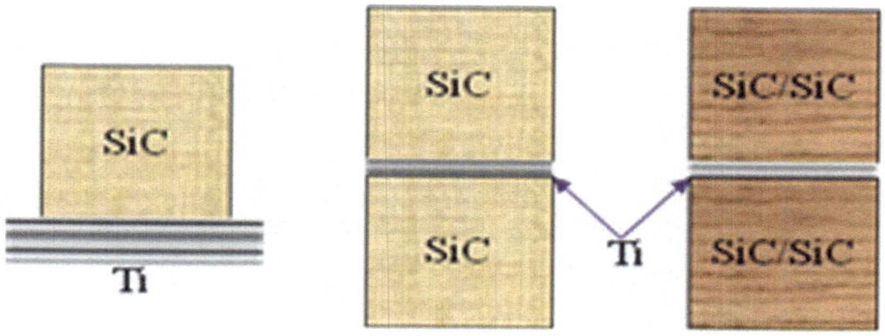

Fig. 1 Schematic diagram of the SiC joining structure

2.2 *Experimental Procedure*

Before joining experiment, the SiC ceramics and SiC_f/SiC specimens were processed into tube shaped of Φ12 (external diameter) × 20 mm and polished to form an annular bonding surface. Then, the Ti is machined to wafer shape in a size of Φ12 × 0.5 mm and applied to the joining surfaces between two SiC ceramics to form the sandwich structure and fixed by a graphite mold at a pressure of 0.5 MPa. Ti is applied to the joining surfaces between two SiC_f/SiC composites and fixed by a graphite mold at a pressure of 0.1 MPa, as shown in Fig. 1. After that, the SiC specimens are put in sintering furnace and heated up to 1500 °C in Ar atmosphere for 60 min, followed by a cooling process to room temperature. Finally, the joined specimens are machined for microstructure analysis.

The microstructure of the joint is observed by optical microscope, and scanning electron microscope (SEM) equipped with an energy dispersive spectrum (EDS). The phase of the joint is detected by X-ray diffractometer (XRD).

3 Results and Discussion

3.1 *Microstructure of SiC Ceramics and SiC_f/SiC Composites*

Microstructure and XRD of SiC ceramics are shown in Fig. 2. The structure of SiC ceramics is compact, and the phase is composed of α-SiC. The density of SiC ceramics is 3.19 g/cm^3 by test. Microstructure and phase of SiC_f/SiC composites are shown in Fig. 3. The structure of SiC_f/SiC composites is relatively compact, and there are a certain number of pores in the composites, and the phase is composed of β-SiC. The density of SiC_f/SiC composites is 2.71 g/cm^3 by test.

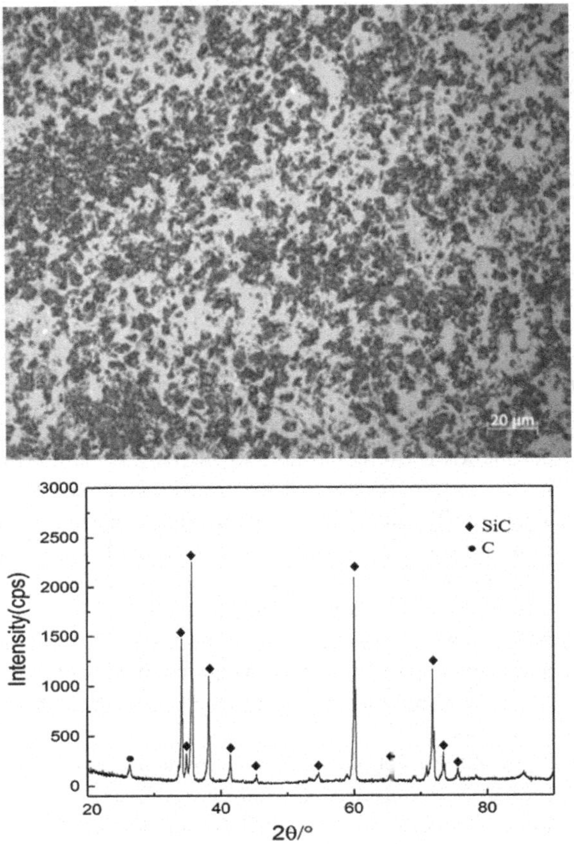

Fig. 2 Microstructure and XRD of SiC ceramics

3.2 Microstructure of the Joint Between SiC Ceramics and Ti

Microstructure of the joint specimens between SiC ceramics and Ti is shown in Fig. 4. A continuous and compact interface layer is formed between SiC ceramics and Ti, the demarcation line between interface layer and SiC ceramics or Ti is clear and smooth.

The analysis of products in interface layer by XRD is shown in Fig. 5. The interface layer is mainly composed of TiC and Ti_5Si_3, and the products are not evenly distributed. The interface layer is mainly composed of TiC and a small amount of Ti_5Si_3 near Ti side, and the content of Ti_5Si_3 is higher than that of TiC near SiC ceramics.

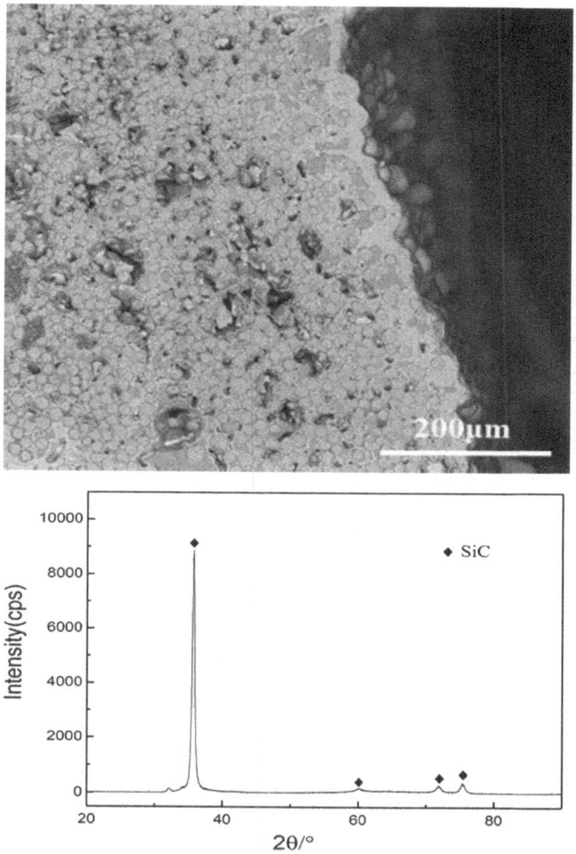

Fig. 3 Microstructure and XRD of SiC_f/SiC composites

The free energy of TiC is lower than that of SiC at high temperature [19], as shown in the following formulas 1 and 2, TiC phase is more stable than SiC. The wetting behavior of metal and ceramics is mainly formed with TiC phase on the surface.

$$\Delta G^0(\mathrm{TiC}) = -183.1 + 0.0101T\ (\mathrm{kJ/mol}) \tag{1}$$

$$\Delta G^0(\mathrm{SiC}) = -53.4 + 0.0069T\ (\mathrm{kJ/mol}) \tag{2}$$

The general process of interfacial reaction between SiC ceramics and Ti is that: (1) free Si and C atoms are produced after the decomposition of SiC ceramics, SiC → Si + C; (2) Ti diffuses into the interface region and reacts with C to generate TiC phase, Ti + C → TiC; (3) Ti continues to diffuse and reacts with the free Si in the interfacial area generating Ti_5Si_3, Ti + Si → Ti_5Si_3.

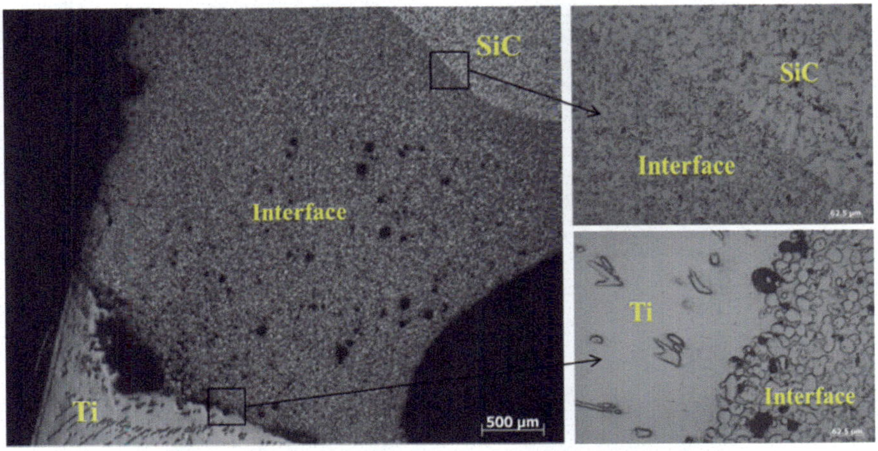

Fig. 4 Microstructure of the joint between SiC ceramics and Ti

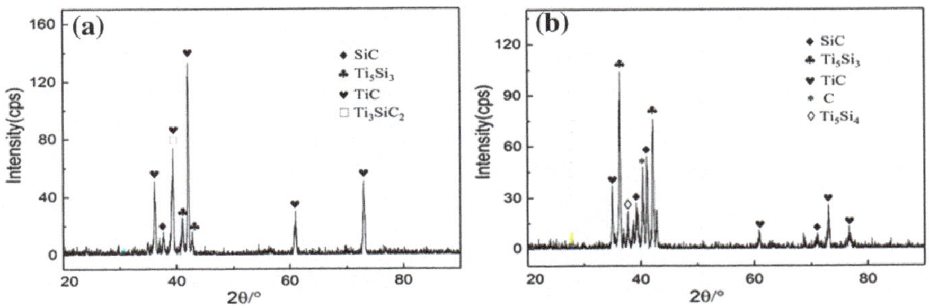

Fig. 5 XRD of interface layer **a** near the Ti side, **b** near the SiC side

3.3 Microstructure of the Joint Between SiC Ceramics

Figure 6a presents joints about 350 μm between SiC ceramics. It shows that the interface layer of Ti between SiC ceramics is completely reacted, composed of a loose middle layer and two compact layers on both sides. The middle layer consists of spherical particles with a small number of pores. The result of EDS line scanning in this area is shown in Fig. 6b. It can be seen that the interface layer is composed of Ti, C and Si elements. The distribution of Ti elements in the interface layer is more uniform. There is a certain degree of diffusion of C and Si elements at the joint layer and content of C elements is higher at the middle interface layer with an obvious enrichment phenomenon, while the diffusion of Si elements is less than that of C in the middle layer because the radius of the Si atom is larger than C that makes the atomic diffusion resistance of Si is large.

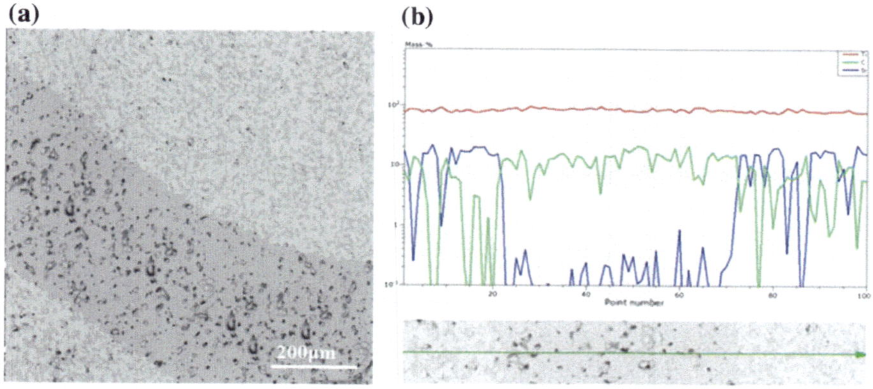

Fig. 6 Microstructure and elemental distribution of the joint between SiC ceramics: **a** morphology; **b** element

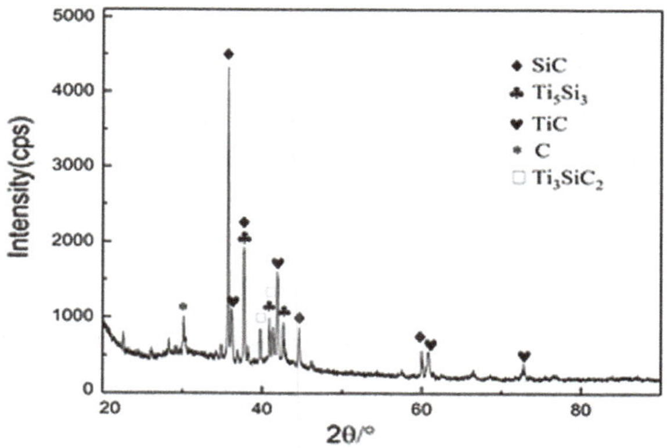

Fig. 7 XRD analysis of the joint between SiC ceramics

The deep color and light color products in the interface layer are analyzed by EDS and XRD (Fig. 7). The joint layer mainly consists of TiC, Ti_5Si_3 and a small amount of Ti_3SiC_2. It is found that the dark phase in the region is composed of Ti and C elements, and the light color phase is mainly composed of Ti and Si elements (Fig. 8). Combined with the results of XRD analysis, the dark color phase is TiC and the light color phase is mainly Ti_5Si_3 and Ti_3SiC_2. The following structure is formed at the joint: SiC/TiC + Ti_5Si_3(Ti_3SiC_2)/TiC/TiC + Ti_5Si_3(Ti_3SiC_2)/SiC.

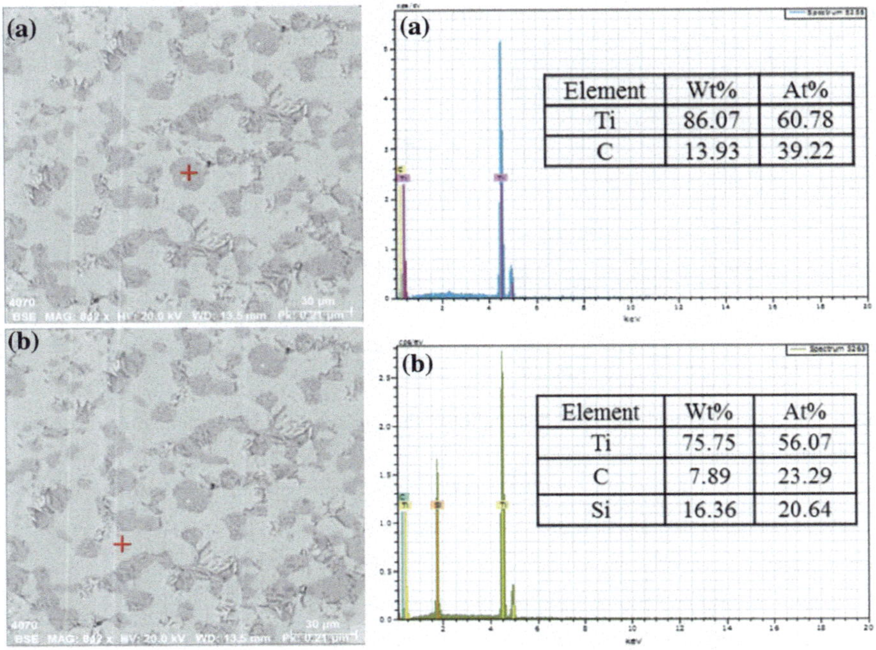

Fig. 8 Elementary composition of the joint between SiC ceramics: **a** dark phase; **b** light phase

3.4 Microstructure of the Joint Between SiC$_f$/SiC Composites

The microstructure of the SiC$_f$/SiC composites joint is shown in Fig. 9. It can be seen that Ti joint between SiC$_f$/SiC composites formed a continuous interface layer with a small number of pores and cracks. The joint has a zigzag structure, which helps to connect the composites. However, a bad contact defect exists in some parts of the interface, which greatly reduces the bonding strength of the joints. In the fiber gap of composites, there are long interfacial products, which penetrate into the interior of the material and formed mechanical occlusion.

The exposed fiber between the interface layer and composites are shown in Fig. 9c. Because of the weak reaction of the SiC fiber, the interface reaction of the SiC matrix first occurs, which leads to the exposure of the fiber and the phenomenon of local fiber bundle bulge. The diffusion of interface reaction to the matrix is restricted by the gap between the prominent fiber and interface layer, which is not conducive to improving the joining performance of the joint. There is a clear and smooth gap between interface layer and composites, when joint surface is parallel to the direction of SiC fiber, as shown in Fig. 9d, there is no obvious mutual diffusion, and the interface with the composite is poor. When the joint is perpendicular to the direction of SiC

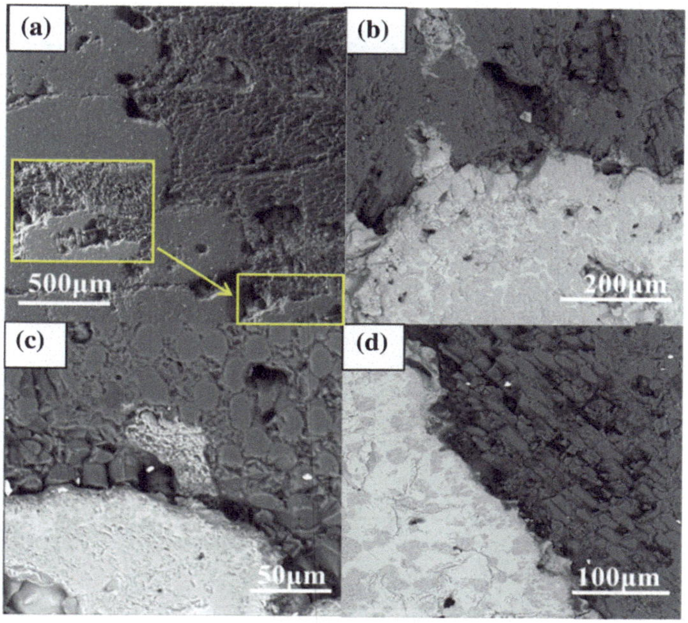

Fig. 9 Microstructure of the joint between SiC_f/SiC composites: **a** microstructure of the composites joint; **b** fiber bundles perpendicular to the joint surface; **c** exposed fibers of the joint; **d** fiber bundles parallel to the joint surface

fiber, interface layer is well combined with the composites, there is obvious mutual diffusion. The interface layer has good contact with SiC fiber and the matrix.

The result of linear SEM-EDS at the joint is shown in Fig. 10a. The violent diffusion reaction results in a large number of C and Si elements in the interface layer. The diffusion capacity of the Si element is weaker than that of the C element, mainly concentrated in the interface layer near the side of the composites, while the Ti element is distributed uniformly in the interface layer, but the diffusion capacity is weak in the composites. The interface layer is detected by XRD (Fig. 10b). Because of the narrow width, phase analysis of interface layer has been greatly affected by the parent material SiC. Besides the SiC phase, there are also TiC, Ti_5Si_3 and Ti_3SiC_2. EDS analyses of dark and light phase in the bonding layer are shown in Fig. 11. It can be seen that dark phase is composed of Ti and C elements, and the light color phase is composed of Ti, Si and C elements. According to the results of XRD detection, the dark phase is TiC, and the light phase is Ti_5Si_3 and Ti_3SiC_2.

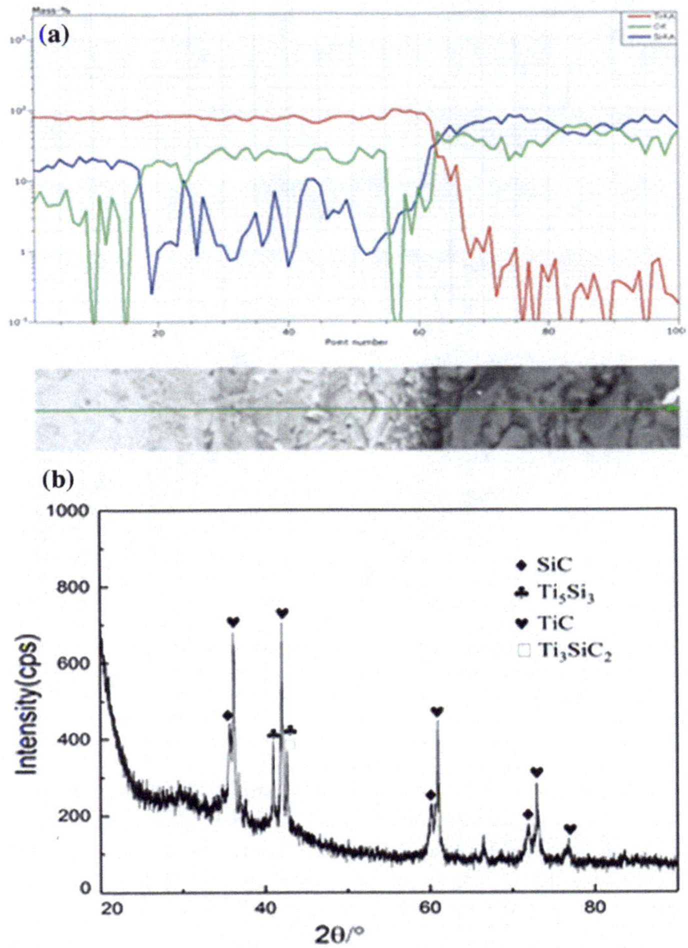

Fig. 10 Elemental distribution and XRD of the joint between SiC_f/SiC composites

The experimental results show that the connection structure of SiC_f/SiC composites is similar to that of SiC ceramics, which forms an interface layer that consists of TiC, Ti_5Si_3 and Ti_3SiC_2. The bonding degree of the composite is not as good as that of the silicon carbide ceramic with the different structure and the arrangement of the SiC fiber which leads to the weak bonding degree of the interface layer.

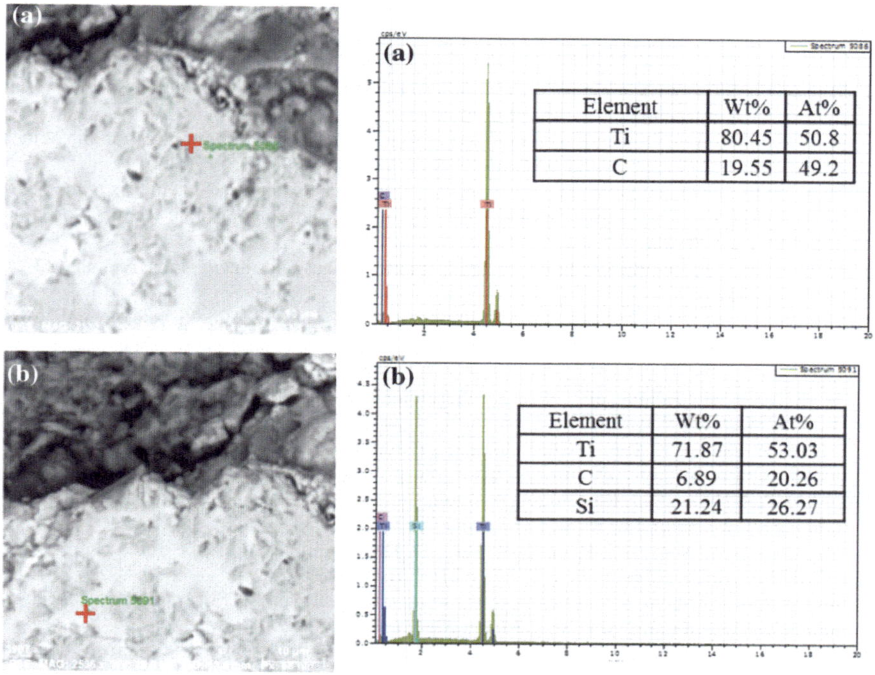

Fig. 11 Elemental consist of bonding layer between SiC_f/SiC composites: **a** dark phase; **b** light phase

4 Conclusion

In this study, the Ti joint has been introduced into the SiC ceramics and 2D Amosic SiC_f/SiC composites by chemical vapor infiltration (CVI) process. SiC ceramics has a good combined effect by strong interdiffusion reaction. The joining has formed a zigzag structure composed of TiC, Ti_5Si_3 and Ti_3SiC_2. Because of the weak reaction capacity with metal, SiC fiber becomes exposed when deep diffusion reactivity takes place in SiC matrix. The diffusion reaction is restricted by the gap between SiC fiber and inter layer, which is not conducive to improving the combined effect. For the combined degree of SiC_f/SiC composites, the joint surface parallel to the direction of SiC fiber is weaker than perpendicular joint.

References

1. R. Naslain, SiC-matrix composites: nonbrittle ceramics for thermo-structural application. Int. J. Appl. Ceram. Technol. **2**(2), 75–84 (2005)
2. R. Naslain, Design, preparation and properties of non-oxide CMCs for application in engines and nuclear reactors: an overview. Compos. Sci. Technol. **64**(2), 155–170 (2004)
3. K. Yoshida, Development of silicon carbide fiber-reinforced silicon carbide matrix composites with high performance based on interfacial and microstructure control. J. Ceram. Soc. Jpn. **118**(1387), 82–90 (2010)
4. L. Shujie, Z. Li, Joining SiC-based materials to themselves and metals. P/M Technol. **22**(2), 91–97 (2004)
5. M. Ferraris, M. Salvo, V. Casalegno, S. Han, Y. Katoh, H.C. Jung, T. Hinoki, A. Kohyama, Joining of SiC-based materials for nuclear energy applications. J. Nucl. Mater. **417**(1–3), 379–382 (2011)
6. M. Salvo et al., Joining of CMCs for thermonuclear fusion applications. J. Nucl. Mater. **1**(233–237), 947–953 (1996)
7. Y. Liu, Z. Huang, X. Liu, J. Chen, Joining strength and microstructure of sintered SiC/SiC joints prepared by active brazing process. J. Inorg. Mater. **24**(2), 297–300 (2009)
8. Y. Liu, Z.R. Huang, X.J. Liu, Joining of sintered silicon carbide using ternary Ag-Cu-Ti active brazing alloy. Ceram. Int. **35**(8), 3479–3484 (2009)
9. H.L. Lee, S.W. Nam, B.S. Hahn, B.H. Park, D. Han, Joining of silicon carbide using MgO-Al_2O_3-SiO_2 filler. J. Mater. Sci. **33**(20), 5007–5014 (1998)
10. M. Ferraris, M. Salvo, V. Casalegno et al., Joining of machined SiC/SiC composites for thermonuclear fusion reactors. J. Nucl. Mater. **375**(3), 410–415 (2008)
11. D. Leeuw, Effects of joining pressure and deformation on the strength and microstructure of diffusion-bonded silicon carbide. J. Am. Ceram. Soc. **75**(3), 725–727 (2010)
12. P. Colombo et al., Joining of reaction-bonded silicon carbide using a preceramic polymer. J. Mater. Sci. **33**(9), 2405–2412 (1998)
13. M.A. Snead, Y. Katoh, T. Koyanagi, G.P. Singh, in *SiC/SiC Cladding Materials Properties Handbook*. United States: N. p., 2017. Web. https://doi.org/10.2172/1399957
14. C.H. Henager Jr., R.J. Kurtz, Low-activation joining of SiC/SiC composites for fusion applications. J. Nucl. Mater. **417**(1–3), 375–378 (2011)
15. Z. Luo, D. Jiang, J. Zhang, Q. Lin, Z. Chen, Z. Huang, Development of SiC-SiC joint by reaction bonding method using SiC/C tapes as the interlayer. J. Eur. Ceram. Soc. **32**(14), 3819–3824 (2012)
16. R. Rosa, P. Veronesi, S. Han et al., Microwave assisted combustion synthesis in the system Ti-Si-C for the joining of SiC: experimental and numerical simulation results. J. Eur. Ceram. Soc. **33**(10), 1707–1719 (2013)
17. Y. Mun, W. Zhou, Y. Hu, Y. Qing, F. Luo, D. Zhu, Improvement of mechanical and dielectric properties of PIP-SiC_f/SiC composites by using Ti_3SiC_2 as inert filler. Ceram. Int. **41**(3), 4199–4206 (2015)
18. C.A. Lewinsohn et al., Silicon carbide-based materials for joining silicon carbide composites for fusion energy applications. J. Nucl. Mater. **307**(2), 1232–1236 (2002)
19. H. Dong, S. Li, Y. He, Joining of reaction bonded SiC ceramic using Ti_3SiC_2 powder as filler. Chin. J. Nonferrous Met. **15**(7), 1051–1056 (2005)

Microstructure and Mechanical Properties of SiBCN Ceramics

Xinyu Xiang, Yunkai Li and Chao Wu

Abstract In this study, four SiBCN ceramic composites (BN+SiC+C, B_4C+Si_3N_4+C, B_2O_3+C+Si_3N_4, B_2O_3+C+Si+AlN) were prepared by mechanical alloying and spark plasma sintering, and their phase, microstructure, open porosity, flexural strength and hardness were characterized. XRD analysis showed that the SiC and BCN were the main phases, in addition to BCN amorphous phase. The B_2O_3+C+Si+AlN composite had the best properties among the four composites, i.e., the porosity as low as 3.67%, the flexural strength of 133.031 MPa, and the Vickers hardness of 4.53 GPa.

1 Introduction

With the continuous development of technology, the requirements of materials' performance in extreme environments (such as ultra-high temperature, ultra-low temperature, ultra-high pressure and extreme vacuum) are becoming more and more demanding. It is important for the mechanical properties of high temperature and high pressure in extreme environments, especially in the aerospace industry [1, 2]. High-temperature heat-resistant structural material has high-temperature resistance, oxidation resistance, ablation resistance and thermal shock resistance. The commonly used high-temperature heat-resistant structural materials are high-temperature alloys, metal-based heat-resistant composite materials and ceramic-based composite materials. However, the use of excellent metal-based heat-resistant composites at a maximum temperature of 1100 °C and the use of silicon-based non-oxide ceramics do not exceed 1600 °C, so it is meaningful to study ceramic materials with excellent high-temperature resistance.

X. Xiang · Y. Li (✉) · C. Wu
Beijing Institute of Technology, Beijing, China
e-mail: liyunkai8866@163.com; liyunkai@bit.edu.cn

X. Xiang
e-mail: Xiangxinyu0827@163.com

Y. Han (ed.), *Physics and Techniques of Ceramic and Polymeric Materials*,
Springer Proceedings in Physics 216,
https://doi.org/10.1007/978-981-13-5947-7_12

SiBCN ceramic is a new type of high-temperature heat-proof structural material [3]. The diffusion rate of O element in SiBCN composite ceramic without grain boundary is not high, so SiBCN composite ceramic has good high-temperature oxidation resistance and work at high-temperature environment of 2000 °C [4] and maintains good high-temperature stability and mechanical properties. In the preparation method of SiBCN ceramics, the mechanical alloying method occupies a pivotal position by virtue of its simple operation, non-toxic raw materials and suitable for mass production. The current research on SiBCN composite ceramic mechanical alloying method is mainly concentrated in the special ceramics research institute of Harbin Institute of Technology. Yang [5] used cubic silicon powder (c-SiC), graphite powder (C) and hexagonal boron nitride powder (h-BN) as raw materials to prepare SiBCN ceramics by high-energy ball milling, hot pressing sintering and spark plasma sintering. In this study, four groups of powders were used to investigate the preparation of SiBCN ceramics by ordinary ball milling and spark plasma sintering.

2 Experimental Part

2.1 Raw Materials and Ratio

By consulting a large number of related literature, we found that amorphous ceramics have the best performance when the atomic ratio Si:B:C:N is 2:1:3:1. So we tried to get close to this ratio and designed four groups of mixed powder. The raw materials used for preparing the SiBCN ceramic powder were BN, SiC, C, B_4C, Si_3N_4, B_2O_3, Si powders, and the Al element was introduced by using AlN powder, and the four groups of SiBCN ceramic were mixed as shown in Table 1.

2.2 Preparation of Massive Ceramics

The mixed raw material powder was ball-milled by a ordinary ball milling with ball-to-batch ratio of 3:1, rotation speed of 300 r/min, ball milling time of 9 h, and then dried and sieved to obtain a dried SiBCN composite ceramic powder. The SiBCN

Table 1 Powder composition mole ratio of SiBCN ceramics (mol%)

Sample	BN	SiC	C	B_4C	Si_3N_4	B_2O_3	Si	AlN
1	25.00	50.00	25.00					
2			72.73	9.09	18.18			
3			69.57		17.39	13.04		
4			46.15			7.69	30.77	15.39

composite powder was sintered by spark plasma sintering. The mold diameter was 20 mm, the sintering temperature was 1670 °C, the sintering atmosphere was vacuum, the sintering pressure was 30 MPa, the holding time was 5 min, and the heating rate was 100 °C/min.

3 Results and Discussion

3.1 Mechanical Properties

The porosity of ceramics has a significant effect on the mechanical properties of ceramics. SiBCN ceramics will produce some pores in the block after SPS sintering, and the pores will lead to stress concentration, which will reduce the mechanical properties (such as strength, flexural strength, and toughness) of the ceramic and cause fracture under low-stress conditions. It is known from Fig. 1a that when the sintering temperature is 1670 °C and the pressure is 30 MPa, the ceramic porosity increases with the increase of the N element content, and the addition of Al element

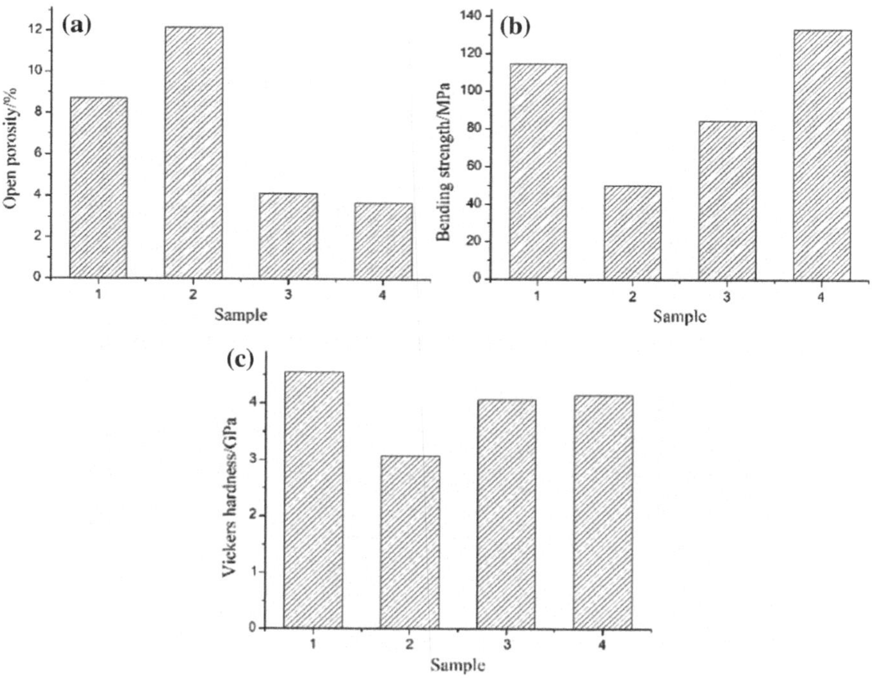

Fig. 1 Mechanical properties of four groups of samples: **a** open porosity; **b** bending strength; **c** Vickers hardness

can moderately reduce the porosity of the ceramic. It can be seen from Fig. 1b that the addition of Al element has a certain effect on the bending strength. The bending strength of samples No. 1 and No. 3 is higher than that of No. 2 sample, and the Vickers hardness of No. 2 sample is also the lowest. This has a certain relationship with their open porosity. The larger the open porosity, the worse the density, and the lower the mechanical properties such as bending strength and hardness. On the other hand, this is related to the performance of their components. B_4C is lower in strength than SiC and the strength of SiC is lower than that of AlN, so the strength of four samples are in the sequence of 1 > 2, 4 > 3.

3.2 Microstructure

It can be seen from Fig. 2 that some of the particulate matter can be observed in samples No. 1 and No. 2, and that there are obvious pores, which leads to an increase in porosity and a decrease in mechanical properties. It is consistent with the results of the test data in the previous chapter. For samples No. 1 and No. 4, it can be seen that the bright SiC phase at the fracture is interlaced with the dark lamellar BCN. The sheet structure in the figure may be the extraction of lamellar

Fig. 2 Fracture SEM photo of four samples: **a** sample No. 1; **b** sample No. 2; **c** sample No. 3; **d** sample No. 4

ceramic particles, the presence of a large number of lamellar ceramic particles on the fracture surface hinders the crack propagation, deflects the crack propagation path, increases the energy required for fracture and crack propagation, thereby increasing the toughness of the ceramic and the mechanical properties of the ceramic.

3.3 Phase Analysis

XRD diffraction analysis of four sets of SiBCN composite ceramics gave the results shown in Fig. 3. Samples 1 and 3 were selected for TEM, HRTEM and SAED analysis to obtain the results shown in Fig. 4. It is known from Fig. 3 that the four ratiometric schemes of the study have amorphous BCN phase formation, indicating that the SiBCN composite ceramic is feasible. The main components of samples No. 1 and No. 3 are BCN phase, 6H-SiC and 3C-SiC. The presence of 6H-SiC indicates that 6H-SiC already exists in the original powder because of the transition temperature of 3C-SiC to 6H-SiC generally higher than 2000 °C. Compared with sample No. 3, Si_3N_4 is also present in sample No. 2. It is preliminarily determined that the increase of B element can promote the decomposition of Si_3N_4.

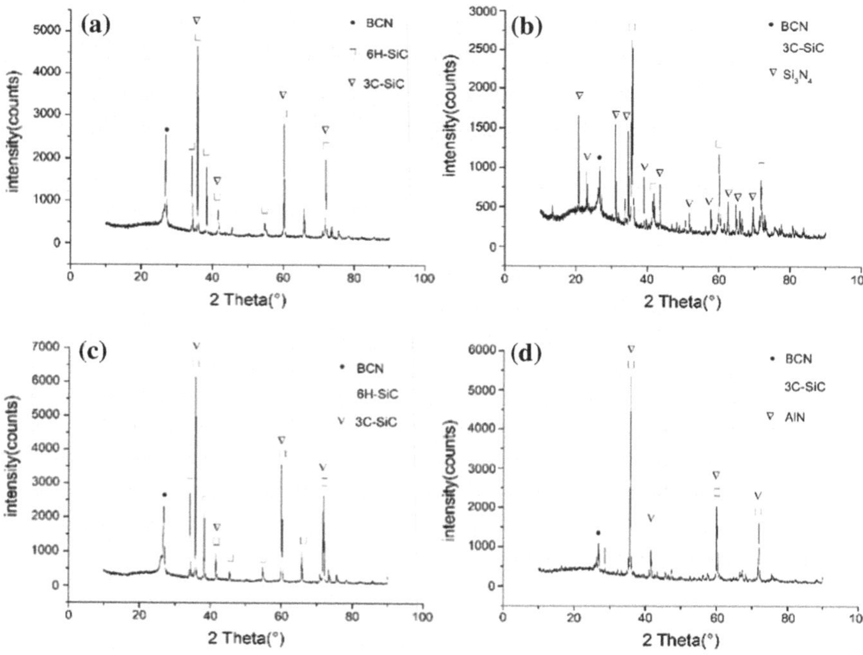

Fig. 3 XRD photo of four samples: **a** sample No. 1; **b** sample No. 2; **c** sample No. 3; **d** sample No. 4

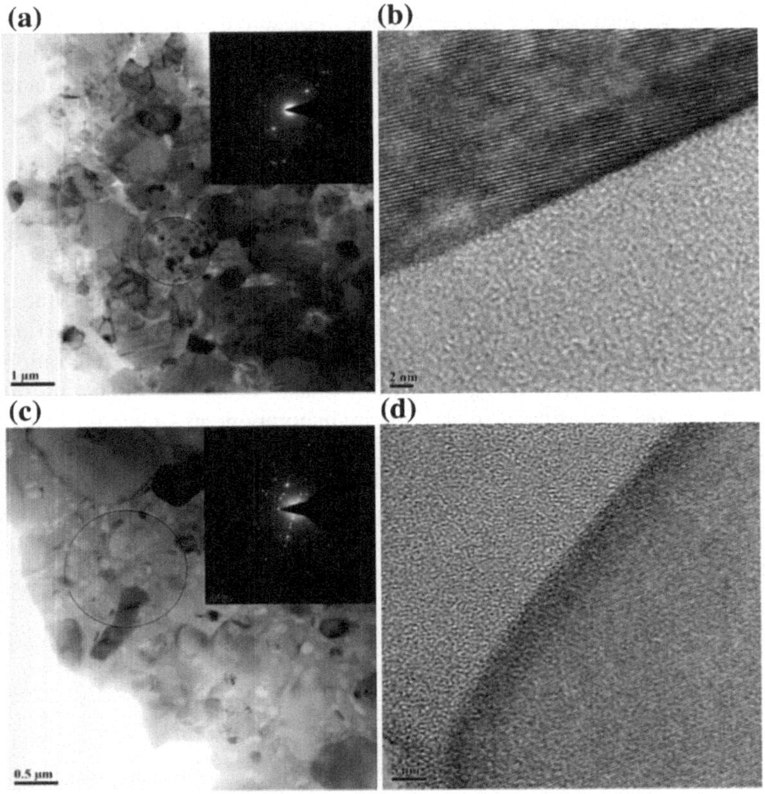

Fig. 4 TEM photo of sample No. 1 and sample No. 3: **a** SAED photo of sample No. 1; **b** HRTEM photo of sample No. 1; **c** SAED photo of sample No. 3; **d** HRTEM photo of sample No. 3

It can be seen from Fig. 4a that the diffraction results of the sample No. 1 are not only concentric rings but also diffuse halo, indicating that the ceramic has a crystalline phase and an amorphous phase. Photographing the interface under high-power transmission electron microscopy showed that there was almost no transition zone between the ordered crystal phase and the diffuse disordered amorphous phase, and the boundary was very obvious. It was proved that the No. 1 sample formed an amorphous phase. The results shown in Fig. 4b are substantially the same as those in Fig. 4a, further illustrating that the sample No. 3 was also a ceramic in which a crystal phase and an amorphous phase coexist.

4 Conclusion

(1) Composite ceramic powders of BN+SiC+C, B_4C+Si_3N_4+C, B_2O_3+C+Si_3N_4, B_2O_3+C+Si+AlN can be prepared into composites containing BCN amorphous phase by ordinary ball milling and SPS sintering;
(2) Among the different ratios of SiBCN ceramics, the composites obtained by the Si:B:C:N ratio approaching 2:1:3:1 has the best mechanical properties;
(3) AlN as an aluminum source adding Al element into SiBCN composite can improve the mechanical properties of ceramics.

References

1. L.F. Jie, G. Li, Z.C. Xu, Hypersonic vehicle concept and development. Flying Missile **5**, 27–31 (2004)
2. K. Upadhya, J.M. Yang, W. Hoffman, Advanced materials for ultrahigh temperature structural application above 2000 °C. Air Force Res. Lab. **76**, 23 (1997)
3. B. Baufeld, H. Gu, J. Bill et al., High temperature deformation of precursor-derived amorphous Si–B–C–N ceramics. J. Eur. Ceram. Soc. **19**(16), 2797–2814 (1999)
4. N.V.R. Kumar, S. Prinz, Y. Cai et al., Crystallization and creep behavior of Si-B-C-N ceramics. Acta Mater. **53**(17), 4567–4578 (2005)
5. Z.H. Yang, in *Microstructure and High-Temperature Properties of the Si-B-C-N MA-Powders and Ceramics* (Harbin Institute of Technology, Harbin, 2008)

Effect of Ambient Temperature on the Emission Spectra of Mg^{2+}- and Ga^{3+}-Doped $CaS:Eu^{2+}$ Red Phosphors

Na Zhang, Renju Cheng, Hanwu Dong, Haili Li, Wenjun Liu, Bin Jiang and Liu Yang

Abstract The photoluminescence spectra of alkaline earth sulfide showed the characteristics of temperature dependence. In this work, Mg^{2+}- and Ga^{3+}-doped $CaS:Eu^{2+}$ phosphors have been synthesized in the tube atmosphere furnace and the influence of ambient temperature on the emission spectra of Mg^{2+}- and Ga^{3+}-doped $CaS:Eu^{2+}$ red phosphors was studied. The results showed that with the increment of ambient temperature, the emission intensities of the $CaS:Eu^{2+}$ phosphors doped with Mg^{2+} and Ga^{3+} decreased rapidly, the emission peaks of the samples changed from deep red to red, and the half peak width increased. This is due to the fact that the high temperature makes more excited molecules at higher vibrational levels, and the fluorescence is mainly caused by the high vibrational energy level of the excited state. By configurational coordinate diagram of the luminescence center, it is confirmed that back tunneling from the ground state to the excited state leads to the increased FWHMs and the decreased emission intensities.

1 Introduction

Most commercial white LED is in the manner of light conversion, which is fabricated by coating YAG phosphors on the chip of blue LED. The YAG phosphors emit yellow light excited by the blue light and the two lights are mixed into white light [1–7]. However, due to the lack of red light components, the light source has high coloring temperature and low color rendering index [8, 9]. The most effective way is to add

N. Zhang · R. Cheng (✉) · H. Dong · H. Li · L. Yang
Chongqing Academy of Science and Technology, Chongqing 401123, People's Republic of China
e-mail: chengrenju@126.com

W. Liu · B. Jiang
College of Materials Science and Engineering, Chongqing University, Chongqing 400045, People's Republic of China

H. Li
College of International Business, Sichuan International Studies University, Chongqing 400031, People's Republic of China

Y. Han (ed.), *Physics and Techniques of Ceramic and Polymeric Materials*, Springer Proceedings in Physics 216,
https://doi.org/10.1007/978-981-13-5947-7_13

red-emitting phosphor on the basis of the blue LED chip [10]. Therefore, there is a demand for novel red phosphor materials that can be effectively stimulated by blue light. Europium-doped alkaline earth sulfides have been considered to be the more promising red phosphor for some time, and their cost of synthesis is very low [11–13].

The synthesis methods of ions as co-doping agents have found a wide utilization in the improving luminescence properties of $CaS:Eu^{2+}$ and these studies have been focused on the persistence or luminescence properties [14–18]. However, there are some problems in this material system for LED light-emitting applications [19, 20]. One of the problems is that $CaS:Eu^{2+}$ has very poor thermostability, that is to say, with the increase of temperature, its photoluminescence intensity decreases rapidly. In previous work [21, 22], by adding small amount of Mg^{2+} and Ga^{3+}, the thermal property of $CaS:Eu^{2+}$ can be improved, and on the basis of previous studies, we have refined the research work.

In this paper, the luminescence intensity and bandwidth of $CaS:Eu^{2+}$, $CaS:Eu^{2+}$, Ga^{3+} and $CaS:Eu^{2+}$, Ga^{3+}, Mg^{2+} at high temperatures are reported. Meanwhile, with the rise of temperature, blue shift is evidently shown in absorption spectra, and an explanation of this phenomenon is given.

2 Experimental

$CaS:Eu^{2+}$ phosphor powders were prepared by using the conversional solid-state reaction method. $CaCO_3$ (analytical grade), S (analytical grade), Eu_2O_3 (99.99%), Ga_2O_3 (99.99%), and $(MgCO_3)_4 \cdot Mg(OH)_5 \cdot 5H_2O$ (analytical grade) were used as raw materials. They were weighted and adequately mixed in the ball mill. Then the homogeneous mixture was fully poured into the corundum crucible. Finally, the mixture was calcined at a certain temperature. The $CaS:Eu^{2+}$ red phosphor powders were obtained. The emission spectra were measured by EX-1000 fluorescence spectrometer. All the tests were carried out from room temperature (RT) to 458 K.

3 Results and Discussion

3.1 *The Effect of Ambient Temperature on Luminescent Properties of the Phosphors*

Figures 1, 2, and 3 show photoluminescence (PL) spectra of $CaS:Eu^{2+}$, $CaS:Eu^{2+}$, Ga^{3+}, and $CaS:Eu^{2+}$, Ga^{3+}, Mg^{2+} under excitation of $\lambda = 460$ nm for various temperatures, respectively. It is shown that in Figs. 1, 2, and 3, the fluorescence emission intensities of three phosphors are decreased significantly, when the testing temperature rises from 298 K to 458 K, the emission maxima of $CaS:Eu^{2+}$ and $CaS:Eu^{2+}$,

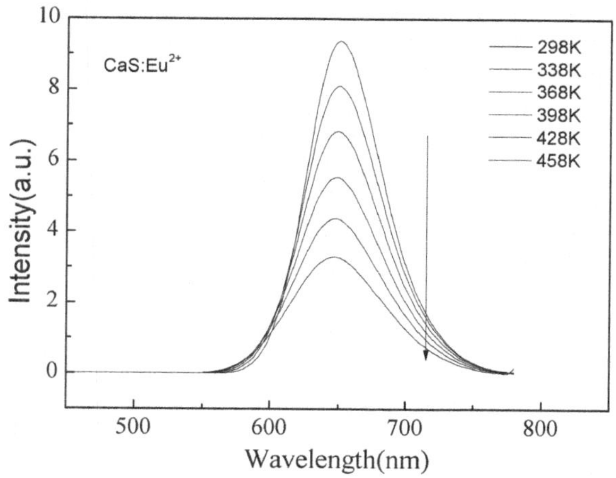

Fig. 1 Emission spectra of $CaS{:}Eu^{2+}$ phosphor under excitation of $\lambda = 460$ nm at different temperatures

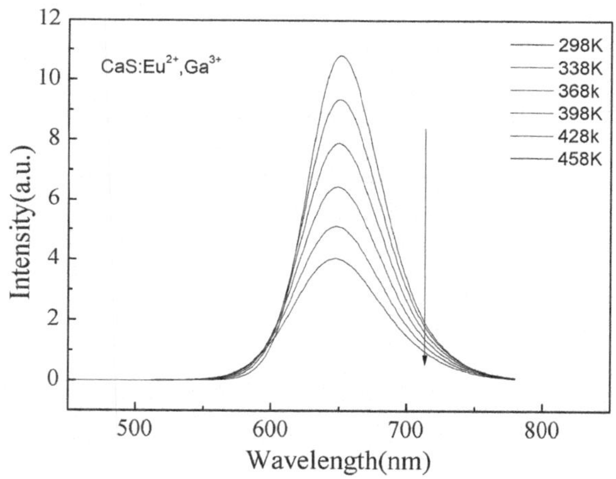

Fig. 2 Emission spectra of $CaS{:}Eu^{2+}$, Ga^{3+} phosphor under excitation of $\lambda =$ 460 nm at different temperatures

Ga^{3+} is moved from 655 to 649 nm, while the emission maxima of $CaS{:}Eu^{2+}$, Ga^{3+}, Mg^{2+} at 653 nm is moved to 649 nm within the same temperature range. The emission maxima of three samples appear blueshift, in other words, the fluorescence spectrum moves toward the short-wave direction of the absorption spectrum, that is, Anti-Stoke's shift. This phenomenon can only be observed at high temperature. This is because that the high temperature makes more excited molecules at higher vibrational levels, and the fluorescence is mainly caused by the high vibrational energy level of the excited state.

The factors affecting the blueshift with the increase of temperature above room temperature are very complicated. This depends on the doping concentrations of divalent europium (Eu^{2+}), the host components and on the temperature ranges investigated. There exists 5d-4f transition in divalent europium ions. In this process, elec-

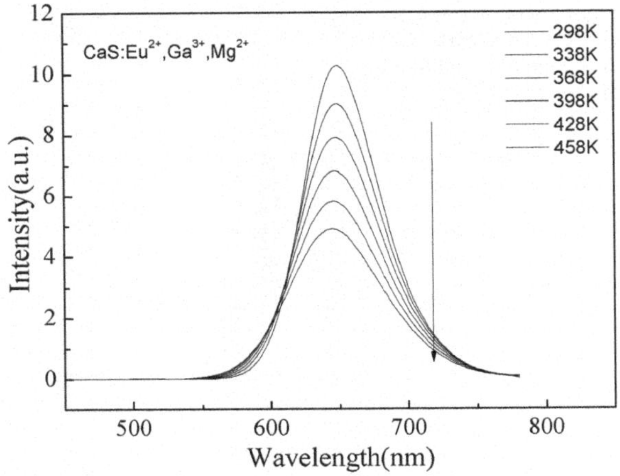

Fig. 3 Emission spectra of CaS:Eu^{2+}, Ga^{3+}, Mg^{2+} phosphor under excitation of $\lambda = 460$ nm at different temperatures

trons in the excited state are unstable and easy to return to the ground state. Energy is lost through radiative and non-radiative transitions. Eu^{2+} interacts with solvents or other molecules to transfer energy resulting in quenching of phosphor. On the other hand, thermal expansion of the host lattice and enhanced interactions between the 5d electrons and phonons at elevated temperatures. They cause blueshifts of the emission maxima of three phosphors. Therefore, as shown in Figs. 1, 2, and 3, the blueshift phenomenon of the emission maxima are observed with the increment of the temperature.

3.2 *The Full Widths at Half Maximum (FWHM) and CIE Chromaticity Coordinates of Phosphors*

FWHM of all samples with increasing temperature is given in Table 1. It can be obtained from Table 1, FWHMs of CaS:Eu^{2+} at 298 K and 458 K are 65.5 nm and 81.5 nm. FWHMs of CaS:Eu^{2+},Ga^{3+} at 298 K and 458 K are 64.6 nm and 79 nm, meanwhile, FWHMs of CaS:Eu^{2+}, Ga^{3+}, Mg^{2+} in the same temperature range are 65.1 and 79.8 nm. These data indicate that the FWHMS increase with the increase of temperature. This phenomenon can be interpreted using the configurational coordinate diagram of the luminescence center [23–25], where the excited state electrons are thermally activated through interaction with phonons, and the energy of electrons in the form of non-radiation is released from the intersection of the excited state and the ground state. If the ambient temperature goes up, the non-radiative transition ability probability is strong, which results in the increase of FWHM for the emission spectra and the decrease of fluorescence emission intensities in Figs. 1, 2 and 3.

The CIE diagram of three samples at different temperatures is shown in Fig. 4. The CIE coordinates of CaS:Eu^{2+} phosphor are found $X = 0.6999$ and $Y = 0.300$

Table 1 FWHM of three samples under excitation of $\lambda = 460$ nm

Temperature (K)	FWHM (nm)		
	$CaS:Eu^{2+}$	$CaS:Eu^{2+}$, Ga^{3+}	$CaS:Eu^{2+}$, Ga^{3+}, Mg^{2+}
298	65.5	64.6	65.1
338	68.7	67.7	68.5
368	71.5	70.4	71.4
398	74.5	73.2	74.3
428	77.7	76.4	77.0
458	81.5	79.0	79.8

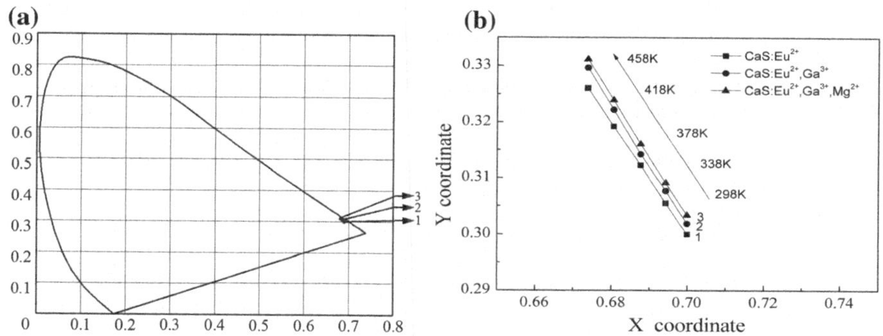

Fig. 4 CIE coordinates of three samples excited at 460 nm at different temperatures: **a** scale drawing; **b** enlarged drawing

at 298 K and $X = 0.6738$ and $Y = 0.3260$ at 458 K. The CIE coordinates X and Y of $CaS:Eu^{2+}$, Ga^{3+} phosphor have value 0.6999 and 0.3019 at 298 K and 0.6738 and 0.3296 at 458 K, respectively. Meanwhile, the CIE coordinates of $CaS:Eu^{2+}$, Ga^{3+}, Mg^{2+} phosphor have value (0.6999, 0.3296) at 298 K and (0.6738, 0.3311) at 458 K. These indicate that the CIE coordinates of the phosphors shift significantly from deep red region to red region. So, as the temperature grows, the blueshift of emission spectra supports the change of CIE coordinates.

4 Conclusions

Mg^{2+}-doped and Ga^{3+}-doped $CaS:Eu^{2+}$ phosphors have been sintered in an atmosphere furnace under Ar/H_2 and their luminescent characteristics for various temperatures are investigated. The peak intensity shows the abnormal blueshift with increasing the bandwidth and decreasing emission intensity as the temperatures rise. This phenomenon can be explained by the level splitting between ground state and excited state based on the configuration coordination model.

Acknowledgements The authors are grateful to the financial supports from National Natural Science Foundation of China (NSFC) (51504052, 51701035), the foundations from Chongqing Science and Technology Commission (cstc2016jcyjA0157), Chongqing Special Key Technology Innovation of Key Industries (CSTC2017ZDCY-ZDZXX0006).

References

1. S. Ray, P. Tadge, S. Dutta, T.M. Chen, G.B. Nair, S.J. Dhoble, Synthesis, luminescence and application of $BaKYSi_2O_7:Eu^{2+}$: a new blue-emitting Phosphor for near-UV white-light LED. Ceram. Int. **44**(7), 8334–8343 (2018)
2. Y.C. Qiang, Z.F. Pan, X.Y. Ye, M.Z. Liang, J.F. Xu, J.H. Huang, W.X. You, H.L. Yuan, Ce^{3+} doped $BaLu_2Al_4SiO_{12}$: a promising green-emitting phosphor for white LEDs. J. Lumin. **203**, 609–615 (2018)
3. A. Tang, D.F. Zhang, L. Yang, Photoluminescence characterization of a novel red-emitting phosphor $In_2(MoO_4)_3:Eu^{3+}$ for white light emitting diodes. J. Lumin. **132**(6), 1489–1492 (2012)
4. Y. Zhou, J. Liu, X. Yang, X. Yu, J. Zhuang, A promising deep red phosphor $AgLaMo_2O_8:Pr^{3+}$ with Blue excitation for white LED application. J. Electrochem. Soc. **157**(3), H278–H280 (2009)
5. H.D. Ju, X.J. Deng, Z.H. Weng, Q. Wu, B. Wang, Y.H. Ma, H. Wang, $LiSr_3SiO_4Cl_3$-A novel host lattice for Eu^{2+}-activated luminescent materials. Ceram. Int. **42**(6), 6846–6849 (2016)
6. Q. Liu, J.L. Shen, T. Wu, L.X. Wang, J. Zhang, Q.T. Zhang, L. Zhang, Enhanced luminescence of a Eu^{3+} -activated double perovskite (Na, Li) $LaMgWO_6$ phosphor based on a site inducing energy transfer. Ceram. Int. **42**(12), 13855–13862 (2016)
7. P. Pust, V. Weiler, C. Hecht, A. Tucks, A.S. Wochnik, A.K. Henss, D. Wiechert, C. Scheu, P.J. Schmidt, W. Schnick, Narrow-band red-emitting $Sr[LiAl_3N_4]:Eu^{2+}$ as a next-generation LED-phosphor material. Nat. Mater. **13**(9), 891–896 (2014)
8. A.A. Setlur, W.J. Heward, Y. Gao, A.M. Srivastava, R.G. Chandran, M.V. Shankar, Crystal chemistry and luminescence of Ce^{3+}-doped $Lu_2CaMg_2(Si, Ce)_3O_{12}$ and its use in LED based lighting. Chem. Mater. **18**(14), 3314–3322 (2006)
9. J.C. Sun, X.Y. Mi, L.J. Lei, X.Y. Pan, S.Y. Chen, Z. Wang, Z.H. Bai, X.Y. Zhang, Hydrothermal synthesis and photoluminescence properties of $Ca_9Eu(PO_4)_7$ nanophosphors. CrystEngComm **17**(41), 7888–7895 (2015)
10. Y.D. Huh, J.Y. Park, S.S. Kweon, J.H. Kim, J.G. Kim, Y.R. Do, Phosphor converted three-band white LED. Bull. Korean Chem. Soc. **25**(10), 1585–1588 (2004)
11. W. Lehmann, Alkaline earth sulfide phosphors activated by copper, silver, and gold. J. Electrochem. Soc. **117**(11), 1389–1393 (1970)
12. H. Texin, H. Williams, R. Ward, The effect of activator concentration on the infrared-sensitive Phosphor, Strontium Selenide-Samarium, Europium. J. Am. Chem. Soc. **70**(1), 310–322 (1949)
13. W. Lehamnn, F.M. Ryan, Cathodoluminescence of $CaS:Ce^{3+}$ and $CaS:Eu^{2+}$ phosphors. J. Electrochem. Soc. **118**(3), 477–482 (1971)
14. R.L. Nyenge, H.C. Swart, O.M. Ntwaeaborwa, The influence of substrate temperature and deposition pressure on pulsed laser deposited thin films of $CaS:Eu^{2+}$ phosphors. Physica B **480**, 186–190 (2016)
15. L.X. Wang, Q. Liu, K. Shen, Q.T. Zhang, L. Zhang, B. Song, C.P. Wong, A high quenching content red-emitting phosphor based on double perovskite host $BaLaMgSbO_6$ for white LEDs. J. Alloy. Compd. **696**, 443–449 (2017)
16. C.F. Guo, D.X. Huang, Q. Su, Methods to improve the fluorescence intensity of $CaS:Eu^{2+}$ red-emitting phosphor for white LED. Mater. Sci. Eng., B **130**(1–3), 189–193 (2006)
17. R.L. Nyenge, H.C. Swart, O.M. Ntwaeaborwa, Luminescent properties, intensity degradation and X-ray photoelectron spectroscopy analysis of $CaS:Eu^{2+}$ powder. Opt. Mater. **40**, 68–75 (2015)

18. D.D. Jia, W.Y. Jia, D.R. Evans, W.M. Dennis, H.M. Liu, J. Zhu, W.M. Yen, Trapping processes in CaS:Eu^{2+}, Tm^{3+}. J. Appl. Phys. **88**(6), 3402–3407 (2000)
19. C.C. Lin, A. Meijerink, R.S. Liu, Critical red components for next-generation white LEDs. J. Phys. Chem. Lett. **7**(3), 495–503 (2016)
20. S. Neeraj, N. Kijima, A.K. Cheetham, Novel red phosphors for solid-state lighting: the system $NaM(WO_4)_{2-x}(MoO_4)_x$:Eu^{3+} (M=Gd, Y, Bi). Chem. Phys. Lett. **387**(1), 2–6 (2004)
21. L. Yang, N. Zhang, R.Y. Zhang, B. Wen, H.L. Li, X.B. Bian, A CaS:Eu based red-emitting phosphor with significantly improved thermal quenching resistance for LED lighting applications. Mater. Lett. **129**(32), 134–136 (2014)
22. N. Zhang, S.Q. Luo, R.J. Cheng, H.L. Li, R.Y. Zhang, B. Jing, L. Yang, in *Temperature Dependence of Ga^{3+}-Doped CaS:Eu^{2+} Phosphors*. Chinese Materials Conference (Springer, Singapore, 2017), pp. 609–615
23. J.S. Kim, Y.H. Park, S.M. Kim, J.C. Choi, H.L. Park, Temperature-dependent emission spectra of M_2SiO_4:Eu^{2+} (M=Ca, Sr, Ba) phosphors for green and greenish white LEDs. Solid State Commun. **133**(7), 445–448 (2005)
24. Y.S. Yan, On the origin of temperature dependence of the emission maxima of Eu^{2+} and Ce^{3+} activated phosphors. Opt. Mater. **79**, 172–185 (2018)
25. S. Shionoya, W.M. Yen, *Phosphor Handbook* (CRC Press, New York, 1998)

Quaternary Ammonium Compounds-Modified Halloysite and Its Antifungal Performance

Xianfeng Yue, Xiaoqing Yang, Huairui Li, Rong Zhang and Daochun Qin

Abstract This paper was aimed to prepare new nanocomposites with antifungal activities by modification of halloysite nanotubes using quaternary ammonium compounds (QACs). Two kinds of QACs (benzyl ammonium chloride and didecyl dimethyl ammonium chloride) were used to modify HNTs by the method of vacuum loading. Fourier Transform infrared spectroscopy and thermogravimetric analysis confirmed that BAC and DDAC have been loaded successfully, and the loading capacity was 9.64 and 11.73%, respectively. The surface characteristic results of zeta potential and water-contact angle analysis showed that modified HNTs had positively charged and hydrophobic surface. The antifungal performance was expressed by inhibition zone test. The result revealed that the HNTs after modified had good inhibition against mould, and the bacteriostasis of DDAC was better than that of BAC. HNTs modified with QACs exhibited great potential applications in the field of antifungal nano-materials and fillers for polymer composites.

1 Introduction

Quaternary ammonium compounds (QACs) are a kind of common cationic surfactant, with halogen anion and long-chain alkyl groups. QACs are well known as environment-friendly bacteriostat, with relatively low toxicity to human beings, as well as good water solubility and poor thermal stability [1]. The main mode of action of QACs against microbial cells is the interaction with cell membranes causing disruption and leakage of the cellular content. The study shows that C12–C14

X. Yue · X. Yang · H. Li · R. Zhang (✉) · D. Qin (✉)
International Centre for Bamboo and Rattan, Beijing 100102, People's Republic of China
e-mail: zhangrong@icbr.ac.cn

D. Qin
e-mail: qindc@icbr.ac.cn

Y. Han (ed.), *Physics and Techniques of Ceramic and Polymeric Materials*,
Springer Proceedings in Physics 216,
https://doi.org/10.1007/978-981-13-5947-7_14

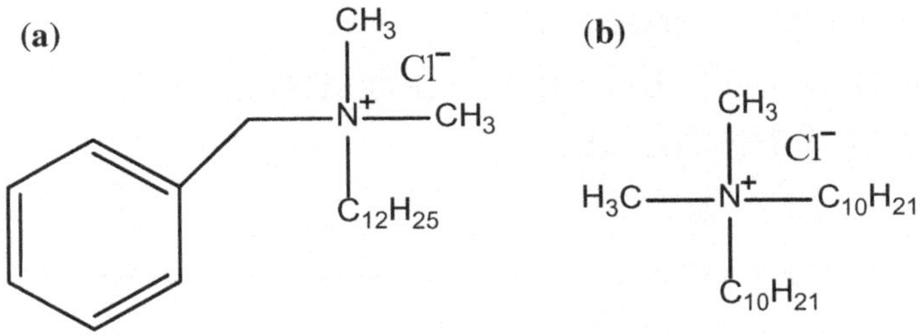

Fig. 1 Chemical structures of the **a** benzyl ammonium chloride (BAC) and **b** didecyl dimethyl ammonium chloride (DDAC)

homologues have the highest biocidal activity against varies microorganisms [2]. QACs were widely used in antibacterial materials [3], disinfection, food hygiene [4], as well as wood preservatives [5].

QACs can be classified into many kinds according to their different structures. Figure 1 shows the general structures of two typical quaternary ammonium compounds. For homologues of the benzyl ammonium chloride (BAC), the cationic nitrogen is surrounded by a benzyl group, two methyl groups and one alkyl chain of chain length *C8–C18,* while for the didecyl dimethyl ammonium chloride (DDAC), it has two alkyl chains of chain length *C10*.

Recently, QACs have been attracting widespread interests in modifying inorganic nanomaterials due to its surfactant abilities. QACs contain hydrophobic alkyl groups and hydrophilic amino group, which are located at both ends of the molecule, forming asymmetric structures. It can be easily attached to inorganic nanomaterials by physical absorption or chemical reaction. Jooyoung (2011) fabricated silica-QAS core–shell NPs with great inhibition activity against the growth of *E. coli*, *S. aureus* and *D. geothermalis* [6]. Montmorillonites (MMTs) can be modified with QACs by the intercalation processes to improve the antibacterial activity. The adsorption of QACs occurred in two steps related to the concentration. At low concentration, the QACs enter the interlayer but do not interact with the outer surface. High concentration allows the QACs enter in a monolayer arrangement and develops electrostatic adsorption mechanisms on the outer surface [7, 8]. Quaternary ammonium groups can also be grafted on the surface of activated carbon for selective adsorption [9].

Halloysite nanotubes (HNTs) are natural aluminosilicate clay with abundant reserves in worldwide. HNTs are tubular nanomaterials with a diameter of 50–80 nm, an inner lumen of 15–20 mm and a length of 600–900 nm [10]. Because of its large aspect ratio, large specific surface area and unique hollow tubular structure, HNTs were widely used in the fields of drug loading and sustained release [11–13], absorption materials [14] and enhancement of composite materials [15]. HNTs have a negative charge on the outer surface and positive charge on the inner surface, so that

HNTs behave unique adsorption properties on both sides of the lumen [11]. After surface modification, HNTs can be well dispersed in the polymer composites and effectively improve its mechanical strength [16]. Wang [17] (2010) modified the surface of HNTs with hexadecyltrimethylammonium bromide (HDTMA) to improve the adsorption capacity of Cr(VI) from wastewaters, and the new adsorbent shows faster adsorption rate and higher adsorption capacity. However, it is not yet known that the surface properties and antifungal performance of QACs-modified HNTs.

In this study, two kinds of QACs (BAC and DDAC) were used to modify halloysite nanotubes via vacuum loading. The QAC is not only filled inside the inner cavity of HNTs, but also modified the external surface of HNTs. The purpose of this study is to impart the antifungal property to HNTs and increase its dispersing performance as well. The QACs/HNTs composites have great potential applications in the field of antifungal coating and polymer fillers.

2 Materials and Methods

2.1 *Materials*

Halloysite nanotubes were obtained from Danjiangkou Mineral Factory (Hubei, China). Benzyl ammonium compounds (BAC) and didecyl dimethyl ammonium chloride (DDAC) were obtained from Aladdin Industrial Corporation (Shanghai, China). Alcohol (95%) was obtained from Beijing Chemical Works (Beijing, China). The experimental strain of *Aspergillus niger V.Tiegh* (As), *Penicillam citrinum Tho* (Pe), *Trichoderma viride Pers.ex Fr* (Tr) and *Botryodiplodia theobromae Pat* (Bt) was obtained from Chinese Academy of Forestry (Beijing, China).

2.2 *Preparation*

HNTs modified with quaternary ammonium compounds (QACs) were prepared in the following way. Firstly, in order to remove the free water, HNTs were dried at 110 °C for 3 h. Then, 2 g of dried HNTs was dispersed in 100 mL saturated solution of QACs (BAC and DDAC). The solvent is water–ethanol mixture (1:1 v/v). The mixture was sonicated for 10–20 min to degas and stirred for 30 min in order to disperse the HNTs. And then the suspension was evacuated by a vacuum pump. The vacuum state was kept for 30 min and then relieved to atmospheric pressure. This vacuum relief process was repeated three times. The loaded composites were obtained by centrifugal precipitation and washed by distilled water for three times to remove excess QACs. Finally, the solid was dried in the oven at 60 °C until constant weight.

2.3 Characterization

Fourier Transform infrared spectroscopy (FT-IR)

Fourier transform infrared spectroscopy (FT-IR) was used to characterize whether the functional groups were grafted onto the surface of HNTs. Samples (about 2% weight) were mixed with KBr and compressed into a thickness of less than 1 mm. Spectra were conducted in between 400 and 4000 cm^{-1} with 4 cm^{-1} resolution and 32 scans by FT-IR spectrophotometer (Nexus 670, Nicolet Instruments, USA).

Thermogravimetric analysis (TGA)

Thermogravimetric analysis (TGA) was used to calculate the loading capacity of QAC on HNTs. TGA was carried out by TA Q 500 (TA Instruments, USA). Samples weighing between 7 and 10 mg were conducted from room temperature to 800 °C at a heating rate of 10 °C/min under nitrogen atmosphere at a flow rate of 40 mL/min.

Zeta potential

Laser particle size analyser (Zetasizer Nano, Malvern Instrument Corporation, UK) was used to measure the zeta potential and to characterize the surface charge changes of halloysites after QAC modification. The sample was dispersed in neutral aqueous solution with a concentration of 0.1%.

Water contact angles

The water contact angles were recorded with OCA 20 (DataPhysics Instruments GmbH, Germany). Samples were pressed into a circular by hydraulic press. The sampling water drops are 5 μL, and the photographic speed is 25 frames per second (frames/s).

Inhibition zone test

The antifungal performance of QACs-modified halloysite was evaluated by inhibition zone test. The experimental method was referred to the *Technical Standard for Disinfection* [18]. The BAC, DDAC, HNTs and QACs-modified HNTs were dispersed in aqueous solution to prepare 1% concentration of QACs. The solution was

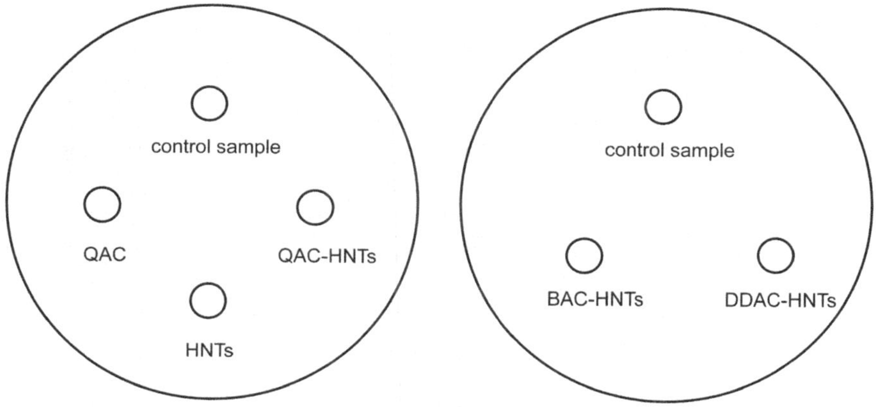

Fig. 2 Diagrammatic sketch of inhibition zone test

dripped about 20 μL onto the sterilized filter paper with a diameter of 5 mm and dried under natural conditions. The prepared hyphae suspension was evenly coated on the PDA medium. As shown in Fig. 2, the filter papers were placed on the medium. Each test condition was repeated three times. The fungal was cultured under the condition of 25–28 °C and 85% humidity for one week. The diameter of each inhibition zone was measured by vernier caliper, and the average value was calculated.

3 Results and Discussion

3.1 *FT-IR*

Figure 3 shows the FT-IR spectra of HNTs samples before and after QACs modification. The original HNTs (Fig. 3a) absorption bands at 3697 and 3621 cm^{-1} were assigned to the stretching vibration due to O–H stretching of the inner surface and inner hydroxyl groups of the HNTs. The absorption band at 1031 cm^{-1} was due to Si–O–Si stretching vibration, and the band at 912 cm^{-1} was attributed to bending vibration of Al–OH. Other bands at 534 and 468 cm^{-1} were attributed to bending vibration and stretching vibration of Si–O, respectively [19, 20]. After the modification of QACs (Fig. 3b, c), it has retained all the original characteristic peaks of HNTs. At the same time, there were two new bands appeared at 2925 and 2854 cm^{-1}, which were attributed to symmetric and asymmetric CH_2-stretching vibration [6]. The presence of these characteristic peaks implies that the QAC has been grafted onto the HNTs surface successfully.

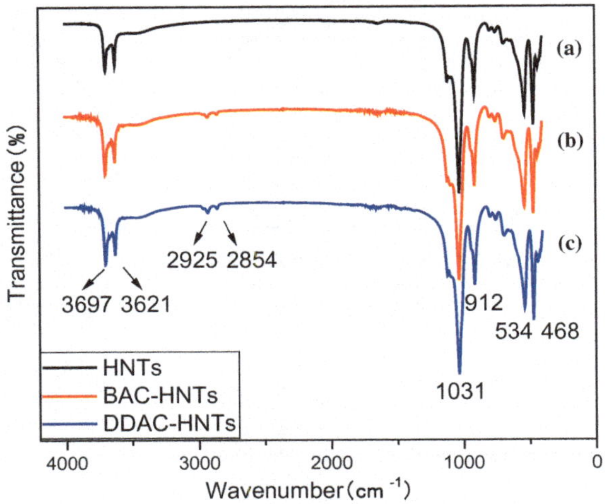

Fig. 3 FTIR spectra of HNTs modified before and after

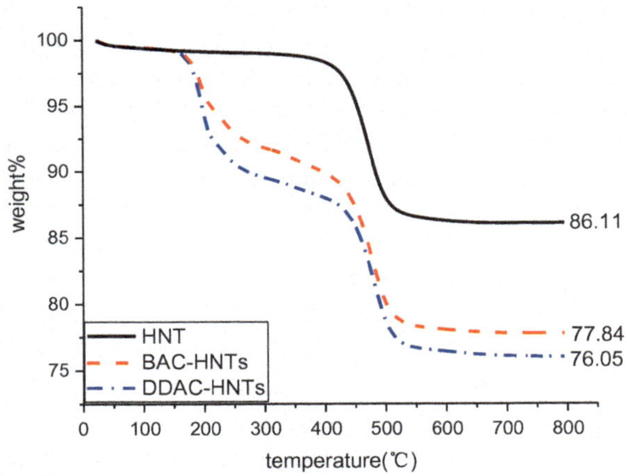

Fig. 4 Thermogravimetric analysis of HNTs modified before and after

3.2 Thermogravimetric Analyses

Different substances have their own decomposition temperatures under the action of heat. Thermogravimetric analyses (TGA) of original and QACs-modified HNTs were used to calculate the loading capacity. Figure 4 shows that the decomposition curves of the QACs-modified HNTs can be divided into three stages. Firstly, the weight loss of HNTs at temperature below 120 °C was caused by the loss of free water. The second stage between 150 and 300 °C was due to the complete decomposition of QACs banded in the HNTs. The last stage between 400 and 550 °C can be attributed to the structural dehydroxylation of Al–OH and Si–OH groups [21].

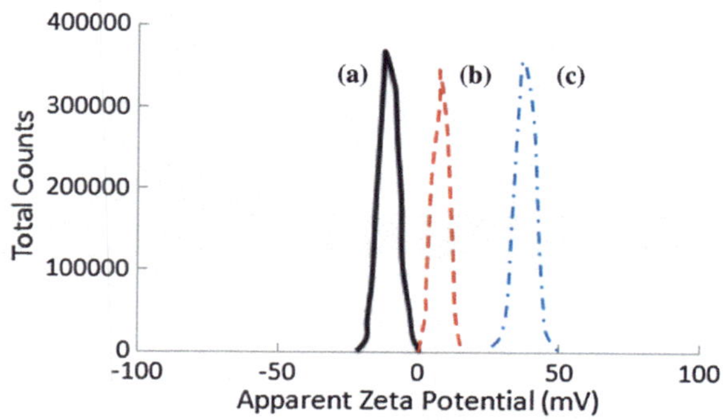

Fig. 5 Zeta potential distribution map of HNTs modified before and after. **a** HNTs. **b** BAC-HNTs. **c** DDAC-HNTs

The loading capacity can be calculated by the following equation. By calculation, the loading capacity of BAC and DDAC on HNTs was 9.64 and 11.73%, respectively.

$$\text{Loading capacity} = \frac{r(HNTs) - r(QACs - HNTs)}{r(HNTs) - r(QACs)} \times 100\% \quad (1)$$

where the r(HNTs), r(QACs-HNTs) and r(QACs) represent the ratio of residual weight of each sample at 800 °C. According to our experiment data, the r(QACs) is about 0.40%.

3.3 *Zeta Potential*

For the original HNTs, the external surface is composed of siloxane (Si–O–Si) groups and silanol (Si–OH) groups [22]. It is usually performed as negative charge, and its zeta potential is measured with the value of −13.7 mV. The surface of halloysite nanotubes can interact with cationic surfactants by electrostatic adsorption. As shown in Fig. 5, after QACs modification, the zeta potential value of BAC-HNTs and DDAC-HNTs was 5.47 and 35.8 mV, respectively. The zeta potential of HNTs after modified by QACs has changed from negative to positive, which indicated that the positively charged quaternary ammonium groups of QACs were covered on the surface of HNTs. DDAC has two long alkyl chains, so the zeta potential of DDAC-HNTs was higher than BAC-HNTs. The zeta potential is a key indicator of the stability of colloidal. For molecules and particles that are small enough, a high zeta potential will confer stability. For further application as nano-filler, DDAC-modified HNTs with a larger zeta potential value will have a better dispersibility.

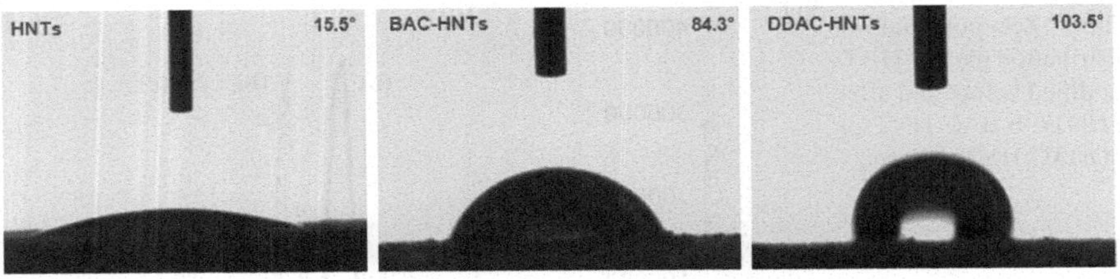

Fig. 6 Water-contact images and numerical contact angle values

3.4 Contact Angle

The contact angle reflects the wettability of the liquid on the solid surface, and the water-contact angle represents the hydrophilic or hydrophobic properties. High specific surface area and the rich hydroxyl groups on the surface of HNTs bring them excellent hydrophilic properties. However, the alkyl long chain of QACs can provide good hydrophobic properties for the surface of HNTs. As shown in Fig. 6, the water-contact angle of original HNTs is measured with the value of 15.5°, while the water-contact angle increased obviously after the QACs modification. QACs have long alkyl groups which provide hydrophobic property to the surface of HNTs. The water-contact angle of DDAC-HNTs (103.5°) is higher than BAC-HNTs (84.3°); this is due to the fact that DDAC has two hydrophobic groups.

3.5 Inhibition Zone Test

Quaternary ammonium compounds are widely used as bacteriostat. Due to the negative electricity of phosphoric acid on the surface of microorganism, QAC can be arranged on the interface of the cell membrane by electrostatic adsorption. The alkyl long chain of QACs can penetrate the cell membrane and then degenerate and inactivate the protein in the cell and destroy the structure of the cell. It can change the permeability of cell membrane, which causes a large number of active substances (such as K^+ and Ribonucleic acid) in the cell to flow out and eventually lead to cell death [23, 24]. The inhibition zone method can qualitatively judge the fungicidal property of the drugs. The inhibition zone diameter reflects the antifungal performance. According to the Chinese Technical Standard for Disinfection [18], when the inhibition zone diameter is more than 7 mm, it is considered to have antifungal activity.

As shown in Fig. 7a, b, the inhibition zone diameter of control samples and HNTs was less than 7 mm. It has been proved that HNTs themselves do not have antifungal activity. The inhibition zone diameters are listed in Table 1. It shows that the inhibition zone diameters of BAC-HNTs and DDAC-HNTs for three kinds of mildew fungi were

(a)
(b)
(c)

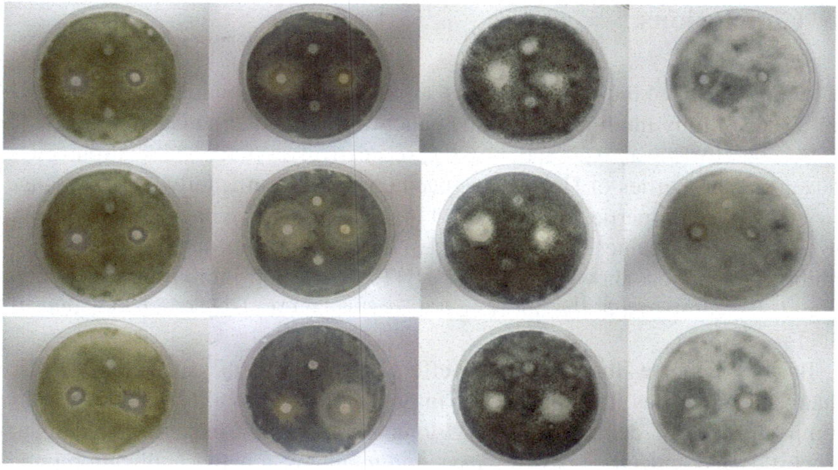

Fig. 7 Inhibition of three kinds of mildew and a stain fungus: **a** inhibition zone of BAC and BAC-HNTs, **b** inhibition zone of DDAC and DDAC-HNTs, **c** inhibition zone of comparison between BAC-HNTs and DDAC-HNTs

Table 1 Result of inhibition zone test

	Inhibition zone (mm)			
	Tr	Pe	As	Bt
Control	*	*	*	*
HNTs	*	*	*	*
BAC	15.12	14.94	13.30	7.53
DDAC	16.47	16.04	13.58	9.35
BAC-HNTs	11.76	12.38	10.56	*
DDAC-HNTs	12.49	14.42	11.03	7.94

*Represents that the diameter of inhibition zone is less than 7 mm

all exceeded 10 mm. However, the inhibition effect on stain fungi was not satisfactory. The modified HNTs put up good antifungal activity, and this can be considered as QACs played a major role in this compound. Compared with QACs, the inhibition zone diameter of QACs-modified HNTs was slightly smaller. This may be due to the QACs loaded in the inner cavity of HNTs released incompletely. Figure 7c shows the comparison of antifungal activity between BAC-HNTs and DDAC-HNTs. The results indicated that the antifungal activity of DDAC was generally stronger than that of QAC. This is attributed to the fact that DDAC has two alkyl long chains, which can provide better antifungal activity.

4 Conclusions

In summary, two kinds of quaternary ammonium compounds (BAC and DDAC) were successfully modified halloysite nanotubes. The loading capacity of BAC and DDAC on HNTs was 9.60 and 11.68%, respectively. After modification of QACs, the surface properties of HNTs have been changed significantly. QACs change the surface charge properties of HNTs and make their surface have good hydrophobic properties. At the same time, QACs still maintain good antifungal performance. The inhibition zone test shows that QACs-HNTs have good inhibition activity against mould. By comparison, the antifungal effect of DDAC-HNTs is better than BAC-HNTs.

Based on the results, QACs can effectively modify the surface of HNTs and change its properties. This may help to improve the dispersibility of HNTs in polymer composite. DDAC contains two alkyl long chains, which make it better in surface properties and antifungal performance, and may have a better application prospect. Meanwhile, the QACs may be loaded in the cavity of halloysite nanotubes, which helps to slow down the release and reduce the loss of QACs.

Acknowledgements We are grateful for the financial support of the Fundamental Research Fund for the International Center for Bamboo and Rattan (No. 1632017020) and the National Key Research and Development Program of China (No. 2017YFD0600803).

References

1. C. Zhang, F. Cui, G. Zeng et al., Quaternary ammonium compounds (QACs): a review on occurrence, fate and toxicity in the environment. Sci. Total Environ. **518–519**, 352–362 (2015)
2. I. Mulder, J. Siemens, V. Sentek et al., Quaternary ammonium compounds in soil: implications for antibiotic resistance development. Rev. Environ. Sci. Biotechnol. **17**, 159–185 (2018)
3. K. Lewis, A.M. Klibanov, Surpassing nature: rational design of sterile-surface materials. Trends Biotechnol **23**(7), 343–348 (2016)
4. M. Krukluoglu, Y. Sahan, A. Yigit, The fungicidal efficacy of various commercial disinfectants used in the food industry. Ann. Microbiol. **56**(4), 325–330 (2006)
5. J. Zabielska-Matejuk, K. Czaczyk, Biodegradation of new quaternary ammonium compounds in treated wood by mould fungi. Wood Sci. Technol. **40**, 461–475 (2006)
6. J. Song, H. Kong, J. Jang, Bacterial adhesion inhibition of the quaternary ammonium functionalized silicananoparticles. Colloids Surf. B **82**, 651–656 (2011)
7. F.M. Flores, E.L. Loveira, F. Yarza et al., Benzalkonium chloride surface adsorption and release by two montmorillonites and their modified organomontmorillonites. Water Air Soil Pollut. **228**(42), 1–14 (2017)
8. K.S. Khalil Rowaida, Selective removal and inactivation of bacteria by nanoparticle composites prepared by surface modification of montmorillonite with quaternary ammonium compounds. World J. Microbiol. Biotechnol. **29**, 1839–1850 (2013)
9. L. Zhang, Y. Liu, S. Wang et al., The removal of sodium dodecyl benzene sulfonate by activated carbon modified with quaternary ammonium from aqueous solution. J. Porous Mater. **24**, 65–73 (2017)
10. P. Yuan, D. Tan, F. Annabi-Bergaya, Properties and applications of halloysite nanotubes: recent research advances and future prospects. Appl. Clay Sci. **112–113**, 75–93 (2015)

11. J. Tully, R. Fakhrullin, Y. Lvov, Halloysite clay nanotube composites with sustained release of chemicals[J]. Nanomater. Nanoarch. **139**, 87–118 (2015)
12. Y. Lvov, W. Wang, L. Zhang et al., Halloysite Clay Nanotubes for Loading and Sustained Release of Functional Compounds[J]. Adv. Mater. **28**(6), 1227–1250 (2015)
13. E. Abdullayev, Y. Lvov, Halloysite clay nanotubes for controlled release of protective age. J. Nanosci. Nanotechnol. **11**, 10007–10026 (2011)
14. J. Matusik, A. Wścisło, Enhanced heavy metal adsorption on functionalized nanotubular halloysite interlayer grafted with aminoalcohols. Appl. Clay Sci. **100**(4), 50–59 (2014)
15. S.M.M. Meira, G. Zehetmeyer, J.O. Werner et al., A novel active packaging material based on starch-halloysite nanocomposites incorporating antimicrobial peptides. Food Hydrocolloids **63**, 561–571 (2017)
16. S. Rooj, A. Das, V. Thakur et al., Preparation and properties of natural nanocomposites based on natural rubber and naturally occurring halloysite nanotubs. Mater. Des. **31**, 2151–2156 (2010)
17. J. Wang, X. Zhang, B. Zhang, Rapid adsorption of Cr (VI) on modified halloysite nanotubes. Desalination **259**, 22–28 (2010)
18. Technical Standard for Disinfection[S]. Beijing: People's Republic of China Ministry of Health (2002), pp. 83–85
19. M. Raji, M.E.M. Mekhzoum, Q.A. el Kacem et al., *Nanoclay modification and functionalization for nanocomposites development: effect on the structural, morphological, mechanical and rheological properties* (Springer Science, Singapore, 2016)
20. P. Krishnaiah, C.T. Ratnam, S. Manickam, Development of silane grafted halloysite nanotube reinforced polylactide nanocomposites for the enhancement of mechanical, thermal and dynamic-mechanical properties. Appl. Clay Sci. **135**, 1–13 (2016)
21. P. Yuan, P.D. Southon, Z. Liu, Functionalization of halloysite clay nanotubes by grafting withγ-aminopropyltriethoxysilane. J. Phys. Chem. C **112**(40), 15742–15751 (2008)
22. L.N. Carli, T.S. Daitx, G.V. Soares, The effects of silane coupling agents on the properties of PHBV/halloysite nanocomposites. Appl. Clay Sci. **87**(4), 311–319 (2013)
23. Y. Zhang, X. Zhao, Research progress in the biocidal performance and mechanism of quaternary ammonium salt biocide. Fine Chem. **27**(12), 1145–1152 (2010)
24. C.J. Ioannou, G.W. Hanlon, S.P. Denyer, Action of disinfectant quaternary ammonium compounds against staphylococcus aureus. Antimicrob. Agents Chemother. **51**(1), 296–306 (2007)

The Development of a New Reinforced Thermoplastic Pipe with Large Diameter for Oil and Gas Transmission Pipeline

Peng Song, Dengzun Yao, Bin Chen and Chao Wang

Abstract Reinforced thermoplastic pipe (RTP) is widely used in petroleum, chemical engineering, and water supply, etc. As RTP is designed to be spoilable, they are not normally larger than 6 in. diameter. However, because of advantage in corrosion resistance, big diameter RTP with high pressure has become the new focus on the field of oil and gas pipeline in the worldwide. In this paper, a new reinforced thermoplastic pipe with a large diameter up to 18 in. for oil and gas transmission pipeline has been successfully developed, with the burst pressure up to 24.34 MPa. The results showed that the failure of glass fiber-reinforced tape was the main fracture mode of this type of RTP under inner pressure. FEA is reliable for RTP design.

1 Introduction

Being corrosion resistant, lightweight, and easy to install at relatively low cost, reinforced thermoplastic pipe is widely used in petroleum, chemical engineering, and water supply, etc., instead of conventional carbon steel pipes. RTP is a composite pipe, which includes a thermoplastic core pipe, an intermediate composite layer, and an outer cover layer, as shown in Fig. 1. The inner and outer layers are made of thermoplastic [usually high-density polyethylene (HDPE)], which are used to containing fluid and providing external protection, respectively. The intermediate layer is made of reinforced tapes, which are a combination of the matrix (usually HDPE) and continuously high-strength fibers (such as carbon, glass, and aramid).

P. Song · D. Yao (✉)
Pipeline Research Institute of CNPC, Langfang 065000, Hebei, China
e-mail: yaodengzun@163.com

B. Chen
The 2nd Pipeline Construction Limited Company of China National Petroleum Corporation, Xuzhou 221008, Jiangsu, China

C. Wang
PetroChina West East Gas Pipeline Company, Pudong 200122, Shanghai, China

Y. Han (ed.), *Physics and Techniques of Ceramic and Polymeric Materials*,
Springer Proceedings in Physics 216,
https://doi.org/10.1007/978-981-13-5947-7_15

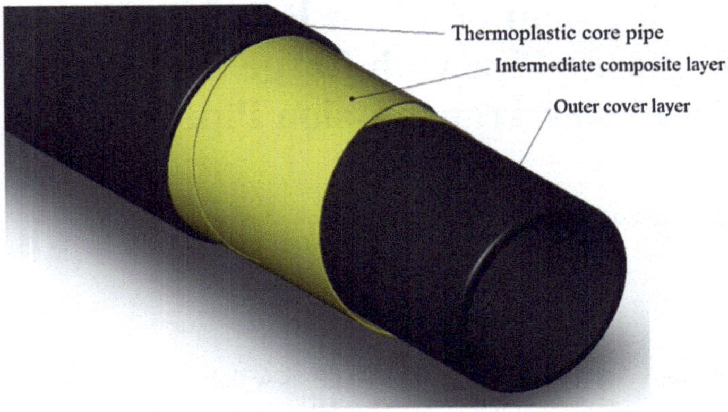

Fig. 1 Typical construction of RTP

And the reinforced tapes are usually twined on the core pipe with a helical winding angle.

As RTP is designed to be spoilable and easy to install from a reel, they are not normally larger than 6 in. (152.4 mm) diameter. For ship transportation, the maximum diameter can be greater up to 9 in. (229 mm) [1]. However, because of it benefit for corrosion resistant, it would be highly advantageous for operators to have RTP as an alternative to steel pipe [2]. And the big diameter RTP with high pressure has become the new focus on the field of oil and gas pipeline in the worldwide.

Now, there are many works have been done for the RTP with the diameter from 2 to 6 in. including the forming process and burst, bend and external pressure collapse behavior of RTP [3–10]. However, there are little reports on the RTP with diameter up to 18 in. In this report, a new reinforced thermoplastic pipe with a large diameter up to 18 in. for oil and gas transmission pipeline has been developed by finite element analysis (FEA) and short-term burst pressure test.

2 Pipe Specifications

In this paper, the 18 in. RTP pipe is developed according to API 15S [11], with the inner and outer layer material is HDPE, while the intermediate layer is made of glass fiber-reinforced tapes with the matrix of HDPE. The reason of using glass fiber-reinforced tapes is for its high tensile strength and relatively low cost, while that for HDPE is because of the environmental friendliness and the compatibility with the glass fiber-reinforced tapes.

According to API 15S, the maximum service pressure (MSP) of RTP can be expressed as:

$$MSP = LCL \times PSF \times F_{cyclic} \times F_{fluid}$$

where MSP is the maximum service pressure; LCL is the lower confidence limit for the product family; PSF is the value of the pressure service factor, as a default value, PSF = 0.67 shall be employed; F_{cyclic} is the cyclic pressure service factor, for static applications, $F_{cyclic} = 1$; F_{fluid} is the fluid service factor, in the absence of any other data, default values of $F_{fluid} = 0.67$ can be used for all hydrocarbon liquid, gas, and multiphase service.

Because this new type pipe is designed as the alternative to steel pipe in the field of onshore oil and gas transmission pipeline, so the in this case, PSF = 0.67, $F_{cyclic} = 1$, and $F_{fluid} = 0.67$ are used. And the total factor of pipe design shall larger than 0.449. Considering the long-term decline of pipe properties, the total factor of pipe design is chosen as 0.25. To satisfied the application, MSP of this big diameter RTP is set as 6.4 MPa, and the short burst pressure should be not less than 25.6 MPa.

3 Materials Property

It is well known that the properties of the reinforced tape and HDPE are key factors for the RTP design. So in this paper, the glass fiber-reinforced tape and HDPE100, provided by a Chinese company, have been tested firstly. The nominal thickness of the tape used in this paper is 0.3 mm. Figure 2 shows the tensile test property of the HDPE100 and the reinforced tape. For the tape, because of the fluctuation of material property, five samples have been tested, and the average value of the tape tensile strength is 768 MPa, while the Elasticity Modulus and the Poisson ratio are 35.4 and 0.43 GPa, respectively.

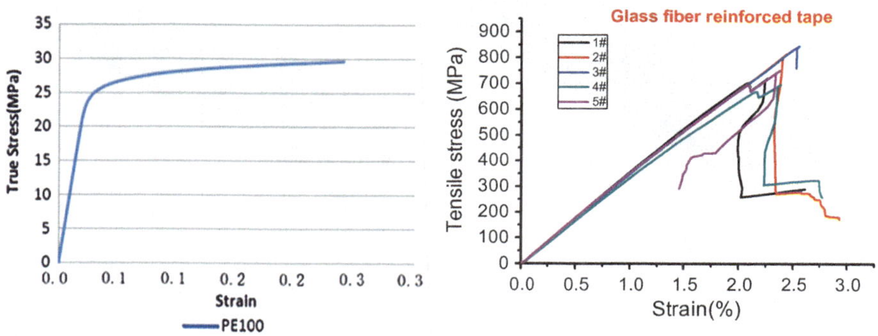

Fig. 2 Tensile test properties of HDPE100 and glass fiber-reinforced tape

4 Finite Element Analyses

The finite element model was developed to predict the short-term burst pressure of RTP pipe. The analysis was performed by employing finite element analysis software ABAQUS. The RTP FEA model and the model mesh are shown in Fig. 3. The model consists of three layers, the inner and outer PE layer, and the intermediate reinforced layer. According to API 15S, the model length of pipe is equal to four times of pipe outer diameter, and the dimension parameters of FEA model are shown in Table 1.

The element of Solid 64 is used for 3D modeling of RTP. The cylindrical coordinates are used in the model, and *R*, *X*, and *Z* indicate the radial, circular, and axial directions, respectively. One of the pipe's end surfaces was fixed in *X*,- *Y*-, and *Z*-direction, representing that the end was totally fixed by the end fitting. Another end surface was fixed in *X*- and *Y*-direction, providing an axially free boundary condition. Internal pressure *P* was applied on the inner face of the pipe, and pipe edge load *F* was applied on one end of model. Model edge load could be determined by

$$F = \frac{2PR_0^2}{R_0 + R_i} \tag{1}$$

To obtain short-term burst pressure, the maximum strain failure criteria were employed to determine the failure initiation of ABAQUS model. It was considered that once the glass fiber-reinforced tape and HDPE reach their strength limits, the RTP pipe fails. The FEA results of intermediate layer under internal pressure are represented in Figs. 4 and 5 for 34 layers and 30 layers, respectively.

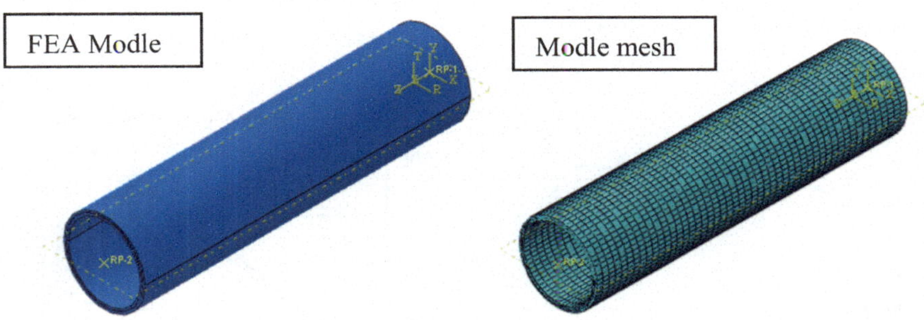

Fig. 3 RTP FEA model and the model mesh

Table 1 Dimension parameters of FEA model

Inner diameter (mm)	Thickness (mm)			Layer numbers of Intermediate layer	Layer angle of Intermediate layer	Pipe length (mm)
	Inner layer	Intermediate layer	Outer layer			
450	10	9, 10.2	5	30, 34	55°	2000

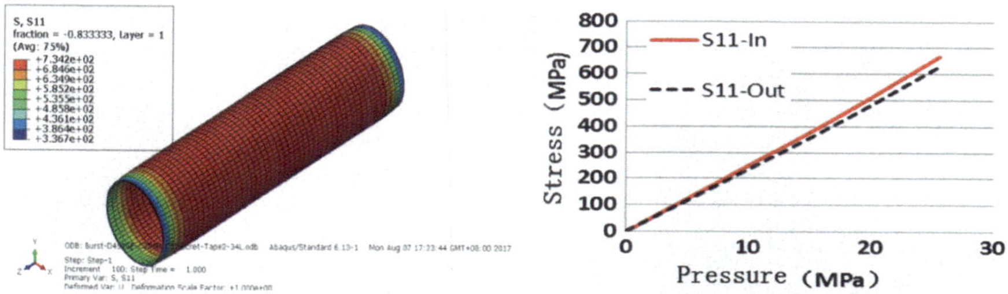

Fig. 4 FEA results of intermediate layer under internal pressure for 34 layers

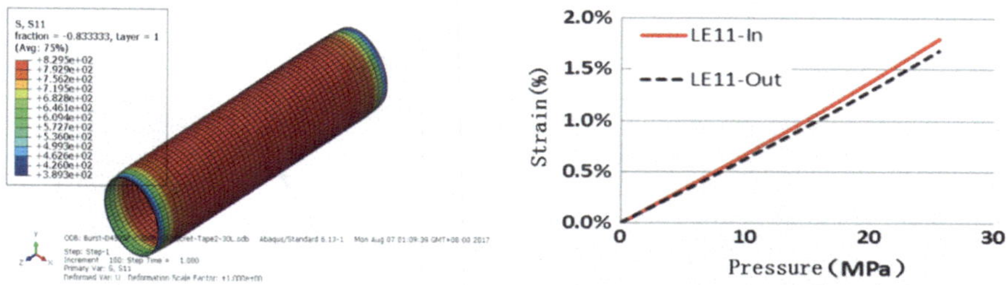

Fig. 5 FEA results of intermediate layer under internal pressure for 30 layers

From Fig. 4, it can be seen that under the inner pressure of 25.6 MPa, the maximum stress of reinforced tape is 734.2 MPa, which is lower than that of tensile strength of tape (768 MPa). Furthermore, the maximum Mises stress in inner and outer PE layer is 26.8 MPa, which is also lower than that of the test result of PE (27 MPa). So when the number of the intermediate layer is 34, the designed RTP can satisfy the inner pressure of 25.6 MPa.

Figure 5 shows that under the inner pressure of 25.6 MPa, the maximum stress of reinforced tape is 829.5 MPa, which is higher than that of tensile strength of tape (768 MPa). So when the number of the intermediate layer is 30, the designed RTP cannot satisfy the inner pressure of 25.6 MPa.

5 Short-Term Burst Test

Based on the FEA results, 18 in. diameter RTP has been produced in China (as shown in Fig. 6). And Fig. 7 is the short-term burst test results. It can be seen that the average value of burst pressure is 24.34 MPa, which is little lower than the FEA results. The reason of the difference must be induced by the pipe manufacturing process.

Fig. 6 Production of 18 in. diameter RTP

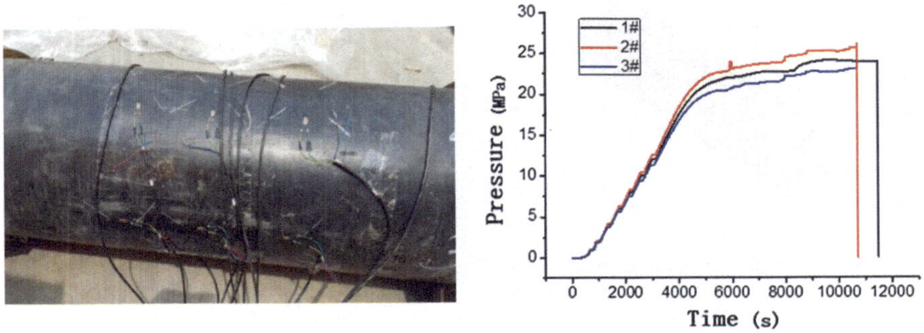

Fig. 7 Short-term burst test of 18 in. diameter RTP

6 Conclusions

In this paper, a new reinforced thermoplastic pipe with a large diameter up to 18 in. for oil and gas transmission pipeline has been successfully developed, with the burst pressure up to 24.34 MPa. It is found that the failure of glass fiber-reinforced tape is the main failure mode of this type of RTP under inner pressure. It also confirms that FEA is reliable for design of RTP.

References

1. F.A. Saleh, I. Ward, Review of flexible composite piping systems, Pipeline Research Council International, Inc., PRCI-473-144506 (2015)
2. I. Ward, F.A. Saleh, State of the art alternatives to steel pipelines, Pipeline Research Council International report PR-473-144506-R01 (2016)
3. Y. Bai, F. Xu, P. Cheng et al., Burst capacity of reinforced thermoplastic pipe under internal pressure. In Proceedings of the ASME 2011 30th International Conference on Ocean, vol. 4 (2011), pp. 281–288
4. Y. Bai, B. Yu, P. Chen et al., Bending behavior of reinforced thermoplastic pipe. J. Offshore Mech. Arctic Eng. **137**(2), 1–11 (2015)

5. Y. Bai, N. Wang, P. Cheng et al., Collapse of reinforced thermoplastic pipe (RTP) under external pressure, in Proceedings of the ASME 2011 30th International Conference on Ocean (2011)
6. X. Li, J. Zheng, S. Feng, Q. Quan et al., Long-term stress analysis of plastic pipe reinforced by cross-winding steel wire. Press. Vessel Technol. **132**(4), 477–491 (2010)
7. J. Azaiez, K. Chiba, F. Chinesta, A. Poitou, State-of-the-art on numerical simulation of fiber-reinforced thermoplastic forming processes. Arch. Comput. Methods Eng. **9**(2), 141–198 (2002)
8. M. Xia, H. Takayanagi, K. Kemmochi, Analysis of multi-layered filament wound composite pipes under internal pressure. Compos. Struct. **53**(4), 483–491 (2001)
9. A. Onder, O. Sayman, T. Dogan, N. Tarakcioglu, Burst failure load of composite pressure vessels. Compos. Struct. **89**(1), 159–166 (2009)
10. M.P. Kruijer, L.L. Warnet, R. Akkerman, Analysis of the mechanical properties of a reinforced thermoplastic pipe (RTP). Compos. A Appl. Sci. Manuf. **36**(2), 291–300 (2005)
11. American Petroleum Institute, Spoolable reinforced plastic line pipe, API15S (2016)

A Study on the Radial Difference of PLA Monofilament

Jian Lu, Yuewei Liu, Zexu Hu, Hengxue Xiang, Zhe Zhou, Bin Sun, Qilin Wu and Meifang Zhu

Abstract PLA monofilament has great potential for resorbable applications. In this paper, a series of PLA monofilaments with different drawing temperature and drawing ratio were prepared through melt spinning. The crystallization and radial difference of the cross section were carefully studied by DSC and Raman spectra, respectively. The results showed that the crystallinity increased with the increasing cooling time in the cooling process and increased with the decreasing drawing temperature or the increasing drawing ratio in the drawing process. The crystallinity of the core was higher than that of the surface. The molecular orientation of the core decreased with the increasing cooling time and increased with the increasing drawing temperature and the decreasing draw ratio, while there's no significant tendency for the molecular orientation of the surface.

1 Introduction

Poly (lactic acid) (PLA) is one of the most promising synthetic biodegradable polyesters. Compared with other bioresorbable polymers, it has better mechanical strength and thus has potential application in the field of implanting, such as resorbable sutures [1], dental materials, fracture fixations [2, 3] and vascular scaffolds [4–9]. As a one-dimensional structure of PLA material, PLA fiber shows great promise for resorbable applications [10–13] because of its large specific surface area, flexibility and mature production technology with large quantity [14]. It's reported that PLA fibers can be prepared by several different methods, such as melt spinning, electro-spinning and wet spinning [15]. Among these methods, melt spinning is more environmentally friendly and economic. Usually, melt spun PLA fibers are divided into monofilaments and multifilaments. Monofilaments show better biocompatibility

J. Lu · Y. Liu · Z. Hu · H. Xiang · Z. Zhou · B. Sun · Q. Wu · M. Zhu (✉)
State Key Lab for Modification of Chemical Fibers and Polymer Material, College of Materials Science and Engineering, Donghua University, Shanghai 201620, China
e-mail: zmf@dhu.edu.cn

Y. Han (ed.), *Physics and Techniques of Ceramic and Polymeric Materials*, Springer Proceedings in Physics 216,
https://doi.org/10.1007/978-981-13-5947-7_16

than multifilaments. To improve the PLA fiber's application as implant material, it's necessary to study the preparing of PLA monofilament.

Melt spinning of monofilaments has been widely studied [16–21], there's a little difference in preparing between monofilaments and multifilaments. In general, a single fiber of multifilaments is always thinner than the monofilaments, which was caused by the difference of cooling process [17]. After extruded from the spinneret, there's a radial temperature gradient result from the difference of cooling rate: The heat transferred slower among polymers in the fiber than the interface between polymers and flowing cold air, and thus, polymers in the surface are more likely to cool down because it's more close to flowing cold air compared to the core. Accordingly, polymers in the core will stay upon crystallization temperature for more time, which may lead to the higher crystallinity of the core. The difference of crystallization will increase with the increasing fiber diameter. The core and the surface will be divided when the difference is large enough, which destroys the structure of the fiber. To prepare monofilaments with large diameter, faster cooling rate is required after extruded. Water bath is widely used to accelerate the cooling process in industries because of the high thermal conductivity and low cost of water. In this way, the radial difference is reduced, and monofilaments with large diameter are easier to form intact structure.

The radial difference is important not only to the forming of monofilaments (especially monofilaments with large diameter) but also to the mechanical degradation properties. The radial difference of temperature was widely studied by simulation [22]. Ma et al. [21] found the parted core and surface in PLA filament with large diameter (0.08–5 mm) and demonstrated the radial difference. The studies about the radial difference were still few and qualitative. In addition, it's reported the difference of cross section influences the mechanical strength and degradation rate of PLA material: A clinical-grade BRS scaffold, which the core's crystallinity was lower than the surface, lost its mechanical strength faster than expected, because the lower crystallized core degraded faster under the same strain, and thus, the self-catalyzed effect was acuter, accordingly the PLA material degraded faster than expected [23].

In this study, Raman spectrometer was adopted to probe changes of the radial difference of crystallization and molecular orientation during the preparing process. And DSC was adopted as the supplementary for the changes of crystallization. To study the influence of cooling time, a series of as-spun fibers with different cooling time were prepared: We extended the distance between the spinneret and water (hereinafter was donated as the height of air-cooling), and thus, the average cooling rate slowed down, accordingly the cooling time was prolonged. To study the influence of the drawing temperature, a series of fibers with different drawing temperature (70, 75, 80, 85, 90 °C) and same draw ratio (3.5) were prepared and so was to the draw ratio.

2 Experimental

2.1 Materials

PLA chips (Nature Works® 6021D) with glass transition temperature (T_g) of 58.79 °C and melting temperature (T_m) of 166.34 °C was purchased from Nature Works® LLC, MN, USA.

2.2 Preparation of PLA Monofilaments

Melt spinning was carried out in a homemade spinning machine. Prior to processing, the PLA chips were dried to avoid degradation due to hydrolysis and pre-crystallization for feeding during manufacturing in a vacuum oven at 80 °C for 12 h and 100 °C for another 12 h, respectively. In order to obtain monofilament, the dried chips were extruded through a round shape spinneret with one hole (D = 1.0 mm). The melt extrusion temperature was kept at 205 °C and cooling by air and water (ambient temperature), successively. The height of air-cooling was set to 1.5–7.5 cm, respectively (hereinafter the corresponding as-spun fibers were donated as "H1.5," "H3.0," "H4.5," "H6.0," "H7.5").

2.3 Drawing Process

Drawing was carried out for the as-spun fibers by passing fibers around a couple of heated godets, then through another couple of heated godets and finally onto a take-up winder. The first couple of godets was operated at a temperature above T_g as drawing temperature, and the second couple of godets at a temperature near T_c as heat setting temperature. In this paper, H7.5 was selected and drawn at 70–90 °C with draw ratio of 3–5 (hereinafter the corresponding fibers were donated as "70–3.5," "75–3.5," "80–3.5," "85–3.5," "90–3.5," "90–3.0," "90–4.0," "90–4.5," "90–5.0"), the heat setting temperature was set to 110 °C.

2.4 Characterization

2.4.1 Differential Scanning Calorimetry (DSC)

Thermal transitions of monofilaments were analyzed by using TA-Q20 differential scanning calorimeter under nitrogen atmosphere. The DSC scans were performed within the temperature range of 40–200 °C at a heating rate of 10 °C/min. From thermograms, cold crystallization (T_c) and melting (T_m) temperatures were recorded using analysis program available from TA®. The degree of crystallinity (X_c) was estimated considering an ideal melting enthalpy of 93.7 J/g [20] according to the follow equation:

$$X_c = \Delta H_f / 93.7 \times 100\% \quad (1)$$

where the ΔH_f was the melting enthalpy of the sample.

2.4.2 Raman Spectra

The Raman spectra of monofilaments were obtained using in Via Reflex Laser Raman Spectrometer of the wavelength of 633 nm in order to define radial difference of crystallization and molecular orientation. To promise the stability of Raman intensity, the geometric center was taken for the core of cross section, and the edge of the same cross section was taken for the surface. For each sample, three data with similar shape and intensity was taken as valid data.

3 Results and Discussion

3.1 Thermal Properties and Crystallinity

Figure 1 shows the DSC curves of the as-spun fibers prepared with air-cooling of various height and the derived thermal transition temperatures were summarized in Table 1. It can be seen in Table 1 and Fig. 1, there is little difference between as-spun fibers in thermal transitions because of their same component. From Table 1, the T_c of H7.5 was confirmed, so we took 110 °C as heat setting temperature to make fibers fast crystallization after drawing.

From the X_c (%) of Table 1, the degree of crystallinity increased with increasing height of air-cooling except H1.5. Fiber with higher air-cooling needed longer to cool down for the slow heat transfer rate of air compared to water, and thus, this

Table 1 DSC data of as-spun fibers

	T_c (°C)	T_{m1} (°C)	T_{m2} (°C)	ΔH_c (J g^{-1})	ΔH_f (J g^{-1})	X_c (%)
H1.5	111.26	160.41	166.90	33.84	37.56	3.97
H3.0	113.00	160.42	167.19	29.31	32.88	3.81
H4.5	112.19	161.30	166.77	30.28	33.88	3.84
H6.0	111.67	160.64	167.44	29.99	34.47	4.26
H7.5	113.42	160.58	167.22	31.48	36.00	4.82

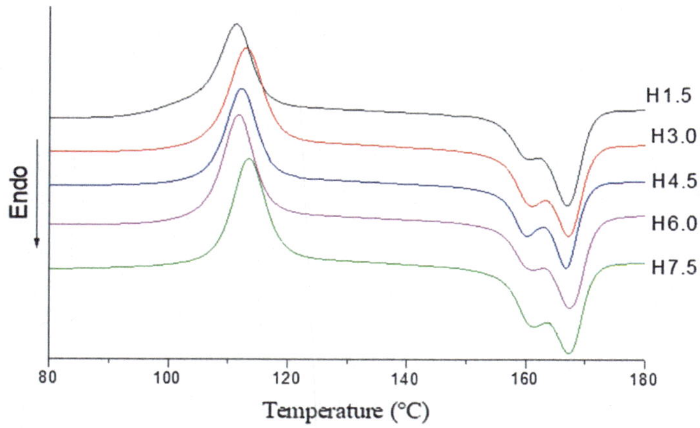

Fig. 1 DSC curves of as-spun fibers

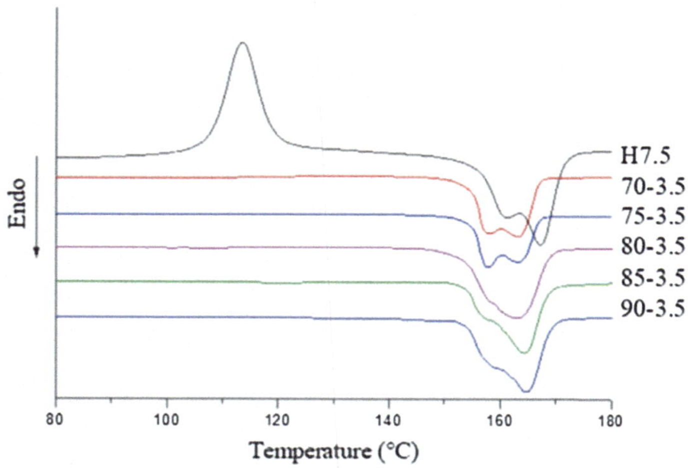

Fig. 2 DSC curves of fibers with different drawing temperature

fiber stayed upon crystallization temperature for longer time, which results in higher crystallinity. For H1.5, it's cooling too fast to reach the diameter of stable winding. And the speed of the fibers was slower than the winding speed because of mass conservation. To reach the winding speed, an accelerated speed was provided by the winding force. And polymers were crystallization to transmit this force to keep force balance inside the fiber.

Figures 2 and 3 show the DSC curves of fibers with various drawing temperature and draw ratio, respectively. Compare Fig. 1 with Figs. 2 and 3, the cold crystallization of 90–130 °C disappeared in fibers after drawing. This can be explained by the drawing process led to perfect crystal structure.

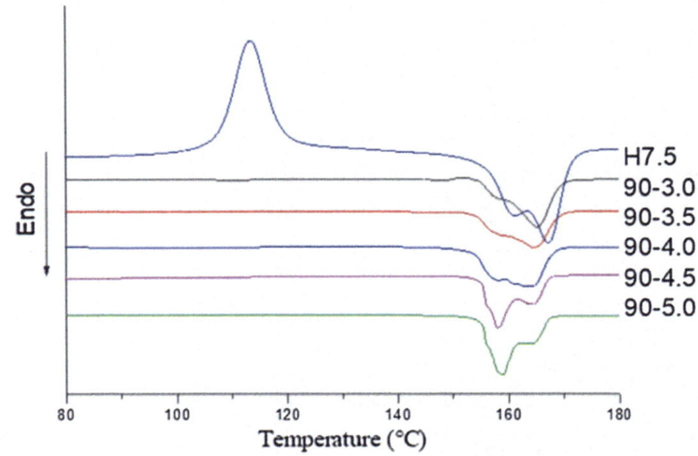

Fig. 3 DSC curves of fibers with different draw ratio

Table 2 DSC data of fibers after drawing

	T_{m1} (°C)	T_{m2} (°C)	ΔH_f (Jg^{-1})	X_c (%)
70–3.5	157.86	163.38	46.21	49.32
75–3.5	157.68	163.43	43.64	46.57
80–3.5	/	164.42	35.86	38.27
85–3.5	/	164.42	33.31	35.55
90–3.5	/	164.74	32.70	34.90
90–3.0	/	165.14	28.82	30.76
90–4.0	158.93	163.74	36.03	38.45
90–4.5	157.99	164.54	39.15	41.78
90–5.0	158.60	164.25	43.51	46.44

From the X_c (%) of Table 2, the degree of crystallinity increased with decreasing drawing temperature and increasing draw ratio. High drawing temperature made molecular active, the more active molecular was, the harder stable structure formed. As a result, less crystal was formed. The force on crystalline polymer would induce polymer chain more active along the force direction. And less energy was required for chain to transform into specific conformation and form crystallization unless the polymer chain became quite straight. As the force increasing with the draw ratio a limited range, crystallinity increased.

3.2 *Radial Difference of Crystallization and Molecular Orientation*

The integrated intensity of the Raman band at 925 cm^{-1} represents the bending mode of CH_3 out-of-plane which was associated with the crystalline phase of PLA [23]. As the crystal increases and becomes preferably aligned in one plane, the bending of CH_3 out-of-plane will result in an amplified signal in response to the incident light, and thus, I_{925} will also be stronger [24].

Figures 4, 5 and 6 showed the Raman spectra of the cross section of prepared fibers, respectively. In Fig. 4, peaks near 925 cm^{-1} of the core could be seen but small in H3.0, H4.5, H6.0 and H7.5, and the intensity of 925 cm^{-1} was weak. This meant the crystallinity of the core was higher than the surface which conformed to the radial difference of crystallization mentioned before. And both of them were so small that the peaks were easy to be interfered.

In Figs. 5 and 6, compared with H7.5, peaks at 925 cm^{-1} of the core and the surface were also small. After drawing, the crystallinity increased on the one hand, these crystals were aligned in the axial direction on the other hand. As a result, the crystals aligned in the plane of the cross section were little, and the intensity of 925 cm^{-1} was weak compared to H7.5. So in this article, peaks at 925 cm^{-1} of prepared fibers were too small compared with peaks at 1452 cm^{-1}, we gave up estimating crystallization with I_{925}/I_{1452} ratio which may lead to inaccurate results.

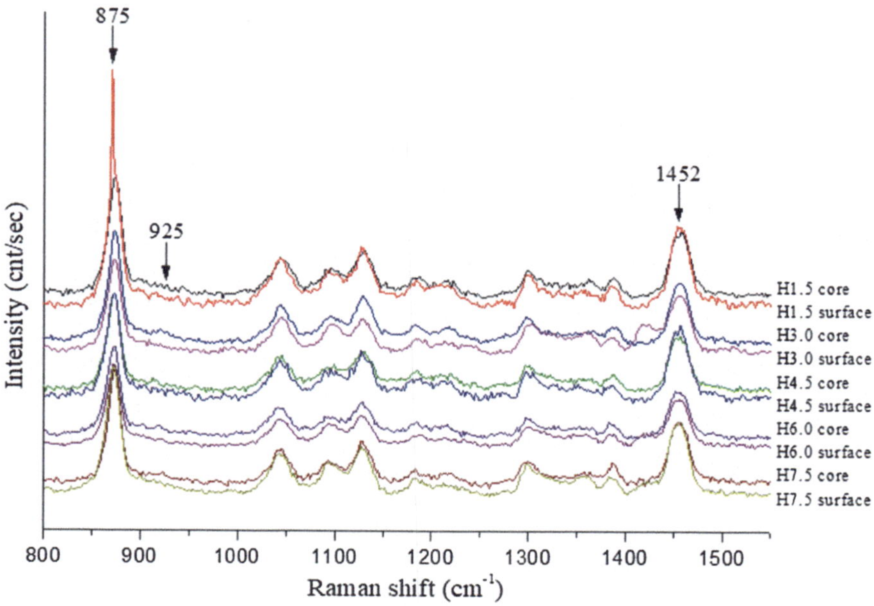

Fig. 4 Raman spectra of the cross section of as-spun fibers

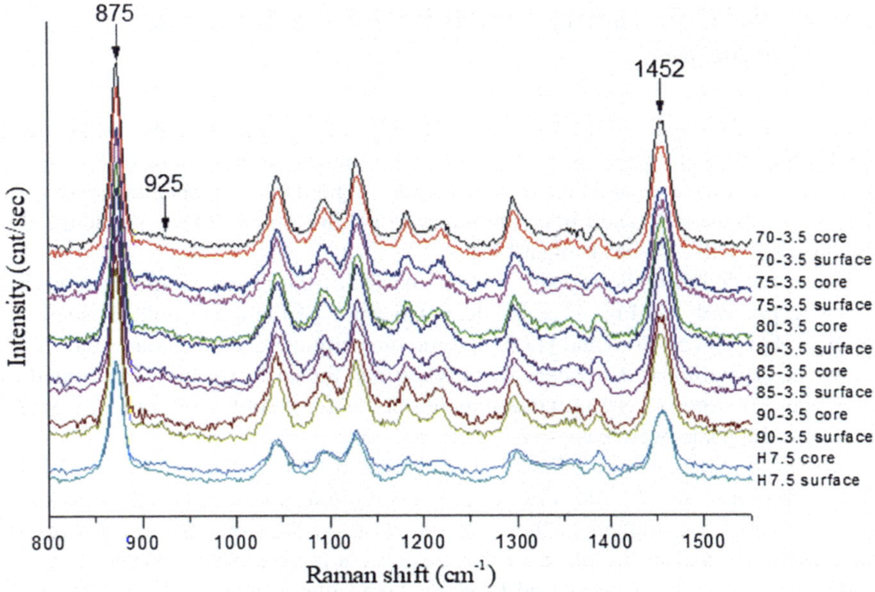

Fig. 5 Raman spectra of the cross section of fibers with different drawing temperature

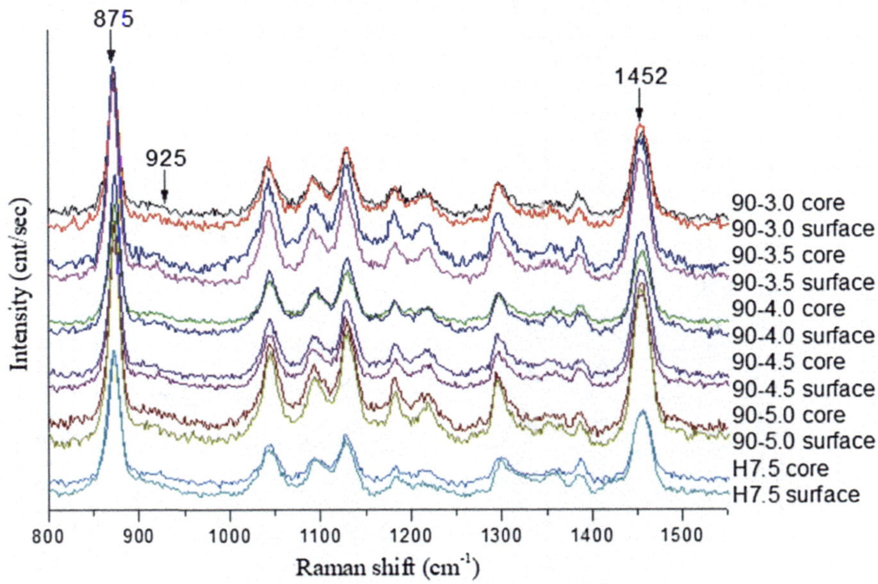

Fig. 6 Raman spectra of the cross section of fibers with different draw ratio

The integrated intensity of the Raman band at 875 cm^{-1} represents the stretch mode of C-COO, the polymer backbone, which was associated with the crystalline phase of PLA [24]. Using the Raman band at 1,452 cm^{-1} (CH3 asymmetric bending mode) taken as an internal normalization standard [25], the I_{875}/I_{1452} ratio can be used as a marker for molecular orientation in one plane [23]. As we all know, the axial direction is perpendicular to the plane of cross section. Without regard to the degradation of polymers, the more polymers aligned in the axial direction, the less polymers aligned in the cross section. So, we can take the I_{875}/I_{1452} ratio of the cross section as a marker for molecular orientation in the axial direction under certain conditions. The radial difference of molecular orientation in the axial direction can be evaluated by measuring the I_{875}/I_{1452} ratio of the different positions on the cross section.

The Raman intensity I_{875}/I_{1452} of the cross section of as-spun fibers with different air-cooling height was summarized in Fig. 7. As can be seen that the intensity radio of the core decreased with the increasing air-cooling height, which is approximate for the surface of H3.0, H4.5, H6.0 and H7.5. Although fibers were drawn in cooling process, the corresponding force was too small. And the polymer chains of as-spun fibers were less oriented in the axial direction so that little difference can be observed for fibers prepared with different cooling height. In addition, as mentioned before, the core of the fiber with higher air-cooling height will stay upon crystallization temperature for more time. High temperature will lead to the degradation of polymer, which results in the cleavage of C-COO and thus destroys the alignment of the polymer chain. Accordingly, the corresponding Raman band intensity (I_{875}) falls. That's why the I_{875}/I_{1452} of the core showed significant reduction with the increasing air-cooling height compared with the surface. It also could be seen that the intensity radio of the core is higher than the surface to as-spun fibers except H7.5, and the

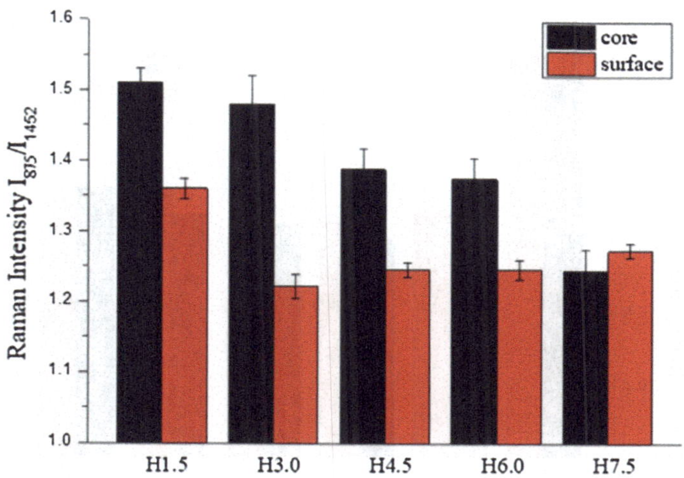

Fig. 7 Raman intensity ratio of the cross section of as-spun fibers

difference between the core and the surface narrowed with the increasing air-cooling height. When the polymer melt flowed steady in the capillary channel of the spinneret, there's radial difference in molecular weight that the molecular weight of the polymer closer to the core tend to be higher, which was determined by the flowing properties of polymer with different molecular weight. As a result, when extruded out of the spinneret, the molecular weight of the core was higher than the surface and so was to the corresponding Raman band intensity. With the cooling time increasing, the molecular weight of the core decreased as mentioned before. As a result, the molecular weight of the core was higher than the surface within the limited range during the cooling process; this is consistent with the tendency shown in Fig. 7. Besides, H7.5 showed the smallest radial difference of molecular orientation, so we took H7.5 as the material to study the influence of drawing temperature and draw ratio.

In Figs. 8 and 9, the Raman intensity I_{875}/I_{1452} of the cross section of fibers with different drawing temperature or draw ratio were summarized and compared with their as-spun fiber (H7.5). It's common that the intensity radio of these fibers were lower than their as-spun fiber. Fibers will become preferably aligned in the axial direction after drawing. And thus the molecular orientation of these fibers will be higher than the as-spun fiber in the axial direction, the molecular orientation in cross section decreased correspondingly. As a result, the corresponding Raman band intensity of the cross section of these fibers became weaker.

In addition, the tendency of the surface's intensity ratio was not obvious with the increasing drawing temperature or draw ratio compared with the core. To prepare fibers with a definite draw ratio, the higher the drawing temperature is, the weaker the force in the axial direction is needed. As mentioned before, for a given as-spun fiber, the crystallinity of the core is higher than the surface. Under the same force,

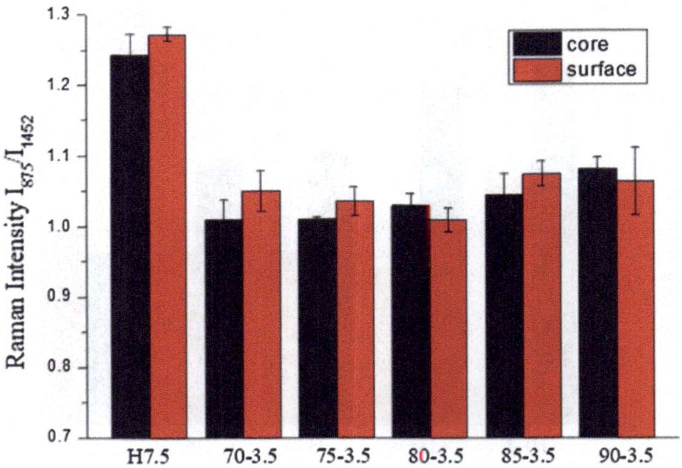

Fig. 8 Raman intensity ratio of the cross section of fibers with different drawing temperature

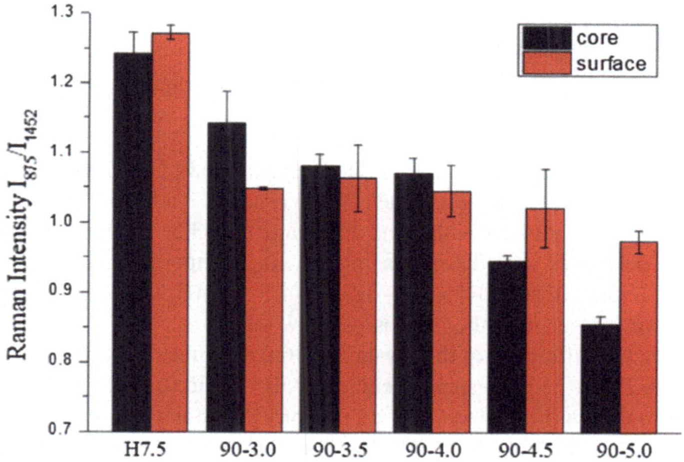

Fig. 9 Raman intensity ratio of the cross section of fibers with different draw ratio

the core with more crystal was harder to be aligned in the direction of the force. Accordingly, the polymer chains in the surface were probably highly aligned in the axial direction during drawing, so that there was no obvious change on the intensity ratio with the increasing drawing temperature or draw ratio. In addition, the polymers in the surface were more exposed to air, which made it easier to degrade compared to polymers in the core. Accordingly, the tendency of the surface's intensity ratio was interfered. Above all, with the increasing drawing temperature or draw ratio, the intensity ratio of the surface showed no obvious tendency.

In Fig. 8, it could be seen that the intensity radio of the core tended to increase with the increasing drawing temperature. Higher the temperature was, more active the molecular of the polymer would be. With a definite draw ratio, the disorientation was stronger with the increasing drawing temperature, while the orientation was still. And thus the polymer chains became less aligned in the axial direction. In other word, the molecular orientation in the cross section was higher, which was more obvious to the core with higher crystallinity. As a result, thus the intensity ratio increased more significant with the increasing drawing temperature.

In Fig. 9, it could be seen that the intensity radio of the core tended to decrease with the increasing draw ratio. The increasing draw ratio made polymer chains easier to be aligned in the axial direction as mentioned before, and thus, polymer chains aligned in the cross section decreased, which led to the weaker intensity of the corresponding Raman band, and thus, the intensity ratio decreased.

4 Conclusions

This study showed that the radial difference of crystallization and molecular orientation depends on the preparing process.

(1) The crystallinity of as-spun fibers (H3.0, H4.5, H6.0, H7.5) increases with the increasing air-cooling height in the cooling process. And in the drawing process, the crystallinity of fibers with same draw ratio (70–3.5, 75–3.5, 80–3.5, 85–3.5, 90–3.5) decreases with the increasing drawing temperature, while at the same drawing temperature (90–3.0, 90–3.5, 90–4.0, 90–4.5, 90–5.0) the crystallinity of fibers increases with the increasing draw ratio.
(2) For the radial difference of the crystallization, the intensity corresponding to the crystal (925 cm^{-1}) in as-spun fibers (H3.0, H4.5, H6.0, H7.5) is stronger in the core than that in the surface, but it's not very obvious because the crystallinity is so small that it's easy to be interfered; the intensity of 925 cm^{-1} is also small because the crystals are more aligned in axial direction after drawing, while the crystals aligned in the cross section become less.
(3) For the molecular orientation, the core's Raman intensity I_{875}/I_{1452} of as-spun fibers (H3.0, H4.5, H6.0, H7.5) decreases with the increasing cooling time because of the longer cooling time led to more degradation of the core; the intensity ratio of fibers after drawing increases with the increasing drawing temperature and the decreasing draw ratio.
(4) By selecting the process method, the radial difference of crystallization and molecular orientation of PLA monofilament could be regulated.

References

1. R.K. Kulkarni, K.C. Pani, C. Neuman, F. Leonard, Polylactic acid for surgical implants. Arch. Surg. **93**, 839–843 (1966)
2. N. Ashammakhi, P. Rokkanen, Absorbable polyglycolide devices in trauma and bone surgery. Biomaterials **18**, 3–9 (1997)
3. I. Bisson, M. Kosinski, S. Ruault, B. Gupta, J. Hilborn, F. Wurm, P. Frey, Acrylic acid grafting and collagen immobilization on poly (ethylene terephthalate) surfaces for adherence and growth of human bladder smooth muscle cells. Biomaterials **23**, 3149–3158 (2002)
4. Y. Wang, X. Zhang, Vascular restoration therapy and bioresorbable vascular scaffold. Regenerative Biomater. **1**(1), 49–55 (2014)
5. Y. Onuma, P.W. Serruys, Bioresorbable scaffold: the advent of a new era in percutaneous coronary and peripheral revascularization. Circulation **123**(7), 79–97 (2011)
6. Y. Wang, L. Kleiner, Fabricating an implantable medical device from an amorphous or very low crystallinity polymer construct. US: 8372332 (2013)
7. Y. Wang, Bioabsorbable stent with layers having different degradation rates. US: 8057876 (2011)
8. Y. Wang, D. Gale, B. Huang, Implantable medical devices fabricated from polymer blends with star-block copolymers. US: 8262723 (2012)
9. Y. Wang, D. Castrol, S. Pacetti, Methods to improve adhesion of polymer coatings over stents. US: 7998524 (2011)

10. W. Channuan, J. Siripitayananon, R. Molloy, G.R. Mitchell, Defining the physical structure and properties in novel monofilaments with potential for use as absorbable surgical sutures based on a lactide containing block terpolymer. Polymer **49**(20), 4433–4445 (2008)
11. S.H. Im, C.Y. Kim, Y. Jung, Y. Jang, S.H. Kim, Biodegradable vascular stents with high tensile and compressive strength: a novel strategy for applying monofilaments via solid-state drawing and shaped-annealing processes. Biomater. Sci. **5**(3), 422–431 (2017)
12. C. Wang, P. Zhang, X. Jiang, Design and characterization of PLLA stents with Z-structure. Text. Res. J. **86**(16), 1701–1709 (2016)
13. E. Tenekecioglu, P.W. Serruys, Y. Onuma, R. Costa, D. Chamie, Y. Sotomi, T.B. Yu, A. Abizaid, H.B. Liew, T. Santoso, Randomized comparison of absorb bioresorbable vascular scaffold and mirage microfiber sirolimus-eluting scaffold using multimodality imaging. Jacc-Cardiovasc. Interv. **10**(11), 1115–1130 (2017)
14. R. Yuval, H.Z. Moran, J.D. Abraham, N. Abraham, Biocompatibility and safety of PLA and its copolymers. Adv. Drug Deliv. Rev. **107**, 153–162 (2016)
15. B. Gupta, N. Revagade, J. Hilborn, Poly (lactic acid) fiber: an overview. Prog. Polym. Sci. **32**(4), 455–482 (2007)
16. R.S. David, *Structure formation in polymeric fibers* (Hanser Publishers, Cincinnati, 2000)
17. H. Ma, Study on the formation machanism and the structure and properties of large-diameter polymer monofilaments. Donghua University (2011)
18. B. Na, N. Tian, R. Lv, S. Zou, W. Xu, Q. Fu, Annealing-induced oriented crystallization and its influence on the mechanical responses in the melt-spun monofilament of Poly(l-lactide). Macromolecules **43**(2), 1156–1158 (2010)
19. S. Ruengdechawiwat, J. Siripitayananon, R. Molloy, R. Somsunan, P.D. Topham, B.J. Tighe, Preparation of a poly(L-lactide-co-caprolactone) copolymer using a novel tin(II) alkoxide initiator and its fiber processing for potential use as an absorbable monofilament surgical suture. International Journal of Polymeric Materials and Polymeric Biomaterials **65**(6), 277–284 (2016)
20. T.H. Oh, Numerical simulation of temperature distribution in melt spinning of PET monofilament. J. Appl. Polym. Sci. **102**(2), 1045–1051 (2006)
21. J. Zhang, H. Ma, H. Zhang, The study on the morphology of large-diameter PLA as-spun monofilaments. Sci. Technol. Inf. **36**, 211 (2013)
22. W.H. Kohler, P. Shrikhande, A.J. McHugh, Modeling Melt Spinning of PLA Fibers. J. Macromolec. Sci. Part B **44**(2), 185–202 (2005)
23. P.J. Wang, N. Ferralis, C. Conway, J.C. Grossman, E.R. Edelman, Strain-induced accelerated asymmetric spatial degradation of polymeric vascular scaffolds. Proc. Natl. Acad. Sci. U.S.A. **115**(11), 2640–2645 (2018)
24. G. Kister, G. Cassanas, M. Vert, B. Pauvert, A. Térol, Vibrational analysis of poly (L-lactic acid). J Raman Spectrosc. **26**, 307–311 (1995)
25. P. Taddei, A. Tinti, G. Fini, Vibrational spectroscopy of polymeric biomaterials. J. Raman Spectrosc. **32**, 619–629 (2001)

Study on Preparation and Properties of Hydrophilic Copolyester of PET-*co*-PEA/Nano SiO_2

Canqing Wu, Xuefeng Mao, Xuzhen Zhang, Chen Lu and Xiuhua Wang

Abstract Polyethylene terephthalate (PET) is the most used fiber material, but its hydrophilic property is weak. Common hydrophilic treatment process on PET clothes has lots of defects such as poor washability and undurablity. In this paper, long-chain monomers were introduced into PET main chains to improve its elastic property to promote the diffusion of water molecules. In addition, modified nano-SiO_2 particles were also added into PET matrix to improve its hydrophilicity. PET-co-PEA/SiO_2 was synthesized by direct esterification with terephthalic acid (PTA), ethylene glycol (EG), polyethylene glycol adipate (PEA), and SiO_2 modified by KH560 on the surface. Morphology, structure, thermal property, and hydrophilicity of PET-co-PEA/SiO_2 were investigated. PET-co-PEA/SiO_2 has a typical PET characteristic absorption peak in FTIR spectra, indicating that the addition of SiO_2 has little effect on PET chemical structure. PET-co-PEA/SiO_2 shows similar thermal stability as PET, which is weaker than PET/SiO_2 but superior to PET-co-PEA. Both nano-SiO_2 particles and the copolymerized long-chain PEA can improve the hydrophilicity of PET. Contact angle of PET-co-PEA/SiO_2 is 69.56°, which is much lower than that of neat PET 108.5°.

1 Introduction

Polyethylene terephthalate (PET) fibers have excellent mechanical properties, such as high modulus, high fracture strength, good dimensional stability, and so on; PET are widely used in clothing, decoration, and industry [1]. However, while PET fiber has many excellent properties, it also has the problems of hygroscopicity and poor dyeing due to its regular molecular structure and lack of polar groups. PET fibers can only be dyed with disperse dyes at high temperature and high pressure or with carrier as the standard moisture return rate is only 0.4% [2–4].

C. Wu · X. Mao · X. Zhang · C. Lu · X. Wang (✉)
National Engineering Laboratory for Textile Fiber Materials and Processing Technology, Zhejiang Sci-Tech University, Hangzhou 310018, Zhejiang, China
e-mail: 15990053712@163.com

Y. Han (ed.), *Physics and Techniques of Ceramic and Polymeric Materials*, Springer Proceedings in Physics 216,
https://doi.org/10.1007/978-981-13-5947-7_17

Owing to lack of moisture absorption and dyeing of PET fiber, the increasing number of research is focused on modification of PET. The main methods of PET fiber modification include chemical method and physical method; one of the chemical modifications is copolymerization modification. Though introducing comonomer into the PET forms new copolyester, these copolyester has some functional aspects, such as dyeability, hygroscopic, antibacterial [5], antistatic [6], flame retardant [7], and so on. Besides, the addition of inorganic nanoparticles in the synthesis, inorganic composite materials can also bring inorganic materials hydrophilic, heat resistance, and other properties into composite materials. At present, the main difficulty of PET modification is how to improve exceptional properties through modification which do not have a great impact on its processability and physical properties. For example, copolymerization modification will greatly affect thermal performance, crystallization performance, and mechanical properties of polyesters. The addition of inorganic nanoparticles will bring adverse effects on the processing process due to aggregation phenomenon. Therefore, according to the processing characteristics PET fiber, the main target is to select the suitable copolymerized monomer, inorganic nano-additive, and inorganic nanoparticle pretreatment process.

Nano-SiO_2 is a common inorganic nano-material particle with a spherical flocculation or reticular structure [8]. In addition, nano-SiO_2 has large surface area, small particle size, and quantum effect of nano-materials. Meanwhile, SiO_2 is widely used in fiber dyeing and hydrophilic modification due to many hydroxyl groups on its surface, and PET/SiO_2 composite have excellent mechanical properties.

In the present paper, nano-SiO_2 was firstly modified by KH-560. Then, PET-*co*-PEA/nano-SiO_2 copolyester was synthesized by direct esterification and melt Polycondensation.

2 Experiment

2.1 Materials

Terephthalic acid (PTA, 99%) was produced by Rongsheng Petrochemical Co., Ltd. (China). Ethylene glycol (EG, 99%), 3-Glycidyloxypropyltrimethoxysilane (KH560, 99%), Nano-silica (SiO_2, 99%), and antimonous oxide (Sb_2O_3, 99%) were purchased from Aladdin Industrial Inc. (China). Polyethylene glycol adipate (PEA, $M_n = 2000$) was supplied by Huafeng group co., Ltd. (China).

2.2 Surface Modification of Nano-SiO_2

The surface modification was carried out in EG medium. SiO_2 was dispersed in EG at a concentration of 1 g/36 ml. KH-560 were then added into the suspension sequentially, and the mount of KH-560 was added at 2 v/v% of EG, followed by ultrasonic treatment for 5.5 h. Schematic of surface modification is shown in Fig. 1.

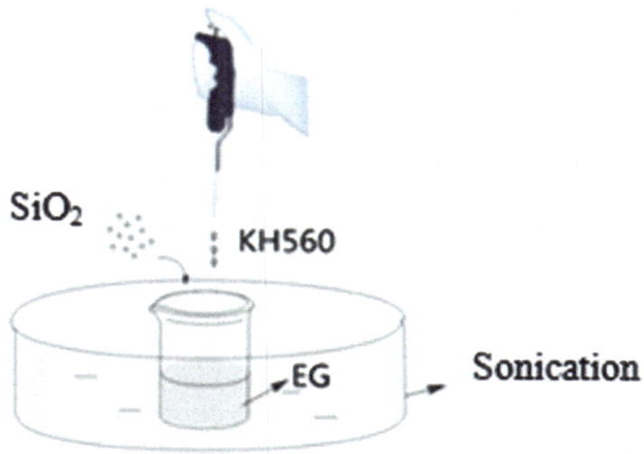

Fig. 1 Surface modification of nano-SiO_2

Table 1 Formulation and intrinsic viscosity of PET-*co*-PEA/nano-SiO_2

Sample	PTA/g	PEA/g	SiO_2/g	EG/ml	Sb_2O_3/g	[η]/dL/g
PET	834	0	0	360	0.3	0.663
PET-*co*-PEA/SiO_2	750.6	72.50	10	360	0.3	0.660
PET/SiO_2	834	0	10	360	0.3	0.671
PET-*co*-PEA	750.6	72.50	0	360	0.3	0.655

2.3 Synthesis of PET-co-PEA/Nano-SiO_2 Copolyester

Esterification reaction: The PTA, PEA, catalytic agent Sb_2O_3, and SiO_2 dispersed in EG medium were added to reaction kettle under nitrogen atmosphere. When the temperature in the reactor reaches 230 °C, the top temperature of the column reaches 100 °C, and the esterification reaction continue 2.5 h. The esterification ends when the top temperature of the column drops back to 100 °C and the water output of esterification reaches 95% of the theoretical value.

Condensation polymerization: After esterification reaction, close the esterification valve and open the polycondensation valve, the temperature inside the kettle rises gradually. Polycondensation was carried out at normal pressure for 0.5 h, and excess EG was steamed out. Then, open the vacuum pump, improve the vacuum in the kettle gradually, when the vacuum degree reaches a higher level, the temperature in the kettle rises rapidly, and the temperature in the reactor is controlled from 270 to 280 °C, the vacuum 100 Pa is maintained, and the rotational speed is adjusted to 60 r/min. When the power is 30 W, the condensation polymerization ends. Technological process is shown in Fig. 2, and formulation and intrinsic viscosity of PET-*co*-PEA/nano-SiO_2 are summarized in Table 1.

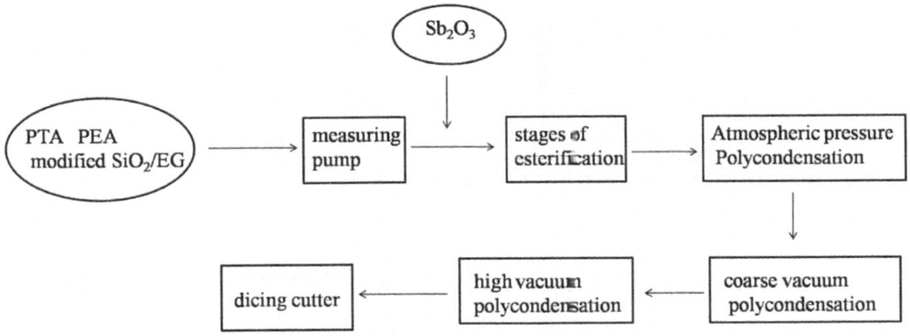

Fig. 2 Flow chart of aggregation process of the PET-*co*-PEA/nano-SiO_2

2.4 Characterization

Morphological analysis of the copolyester specimen was characterized by field emission scanning electron microscopy (FE-SEM) (Carl Zeiss SMT Pte Ltd.) and operating conditions on cryogenically fractured surface. Meanwhile, Morphological of the copolyester specimen was performed on transmission electron microscopy (TEM) (JEM-2100, Japan) operated at an accelerating voltage of 100 keV.

Fourier transform infrared spectroscopy (FTIR) spectrograms were obtained on an FTIR 5700 spectrometer (Thermo Electron Corp., USA). PET copolyester was analyzed in the range of 450–4000 cm^{-1}.

Crystallization behavior of PET copolyesters was tested by differential scanning calorimeter (DSC) instrument (Mettler Toledo Instrument Co., Ltd, Switzerland). Each sample was firstly heated from 25 to 290 °C at a rate of 10 °C/min and maintained at 290 °C for 3 min to erase pervious thermal history. Then, the sample was cooled to 25 °C at a cooling rate of 80 °C/min and maintained 5 min. Lastly, the sample was heated to 290 °C at a rate of 10 °C/min.

Thermal stability of PET copolyesters was tested on Thermogravimetric analysis (TGA) (Mettler Toledo Instrument Co., Ltd, Switzerland). All samples were heated from 25 to 700 °C at a heating rate of 1 0 °C/min under nitrogen atmosphere.

Surface hydrophilicity of PET copolyesters was tested on film surfaces using the OCA40Micro contact angle measurement (Data physics, Germany) at room temperature.

3 Result and Discussion

3.1 *Morphological Analysis*

Figure 3a shows the FE-SEM image of PET-*co*-PEA/nano-SiO_2 copolyester. It can be seen that the size of SiO_2 is about 10–30 nm. No apparent self-aggregation and phase separation are observed on copolyester surface. Meanwhile, Fig. 3b shows that the nano-SiO_2 can disperse homogeneously in PET-*co*-PEA matrix. This phenomenon illustrates that phase interface may have a good compatibility between SiO_2 modified by KH-560 and EG to achieve integration. In addition, PET chains may graft on surface of SiO_2, promote the compatibility between nano-SiO_2 and PET-*co*-PEA matrix.

3.2 *FTIR Investigation*

FTIR spectra of SiO_2 and modified SiO_2 are shown in Fig. 4a. Typically, peaks at 465 cm^{-1} is the bending vibration of Si–O–Si, and peaks at 792 cm^{-1} is symmetric contraction vibration of Si–O–Si. Furthermore, peaks at 3442 cm^{-1} is vibration of the hydrogen bond which is associated by silicone alcohol group [9]. Apparently, modified SiO_2 at 3442 cm^{-1} is significantly weakened compared with origin SiO_2. This result proves that the reaction between KH-560 and SiO_2 resulted in organic surface of SiO_2.

Figure 4b shows that both neat PET and its copolyesters FTIR spectra. The peaks of PET/SiO_2, PET-*co*-PEA, and PET-co-PEA/SiO_2 had no significant change compared with neat PET. The migration peaks at 1718–1720 cm^{-1} may attribute to the

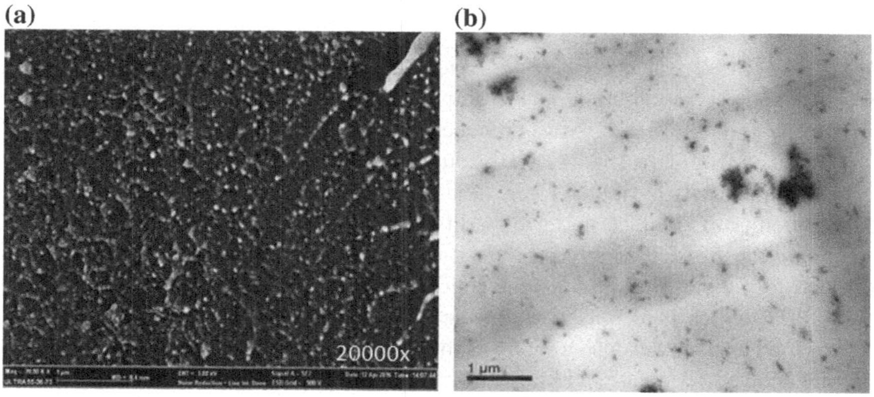

Fig. 3 **a** SEM image of the PET-*co*-PEA/nano-SiO_2 composites and **b** TEM image of PET-*co*-PEA/nano-SiO_2 composites

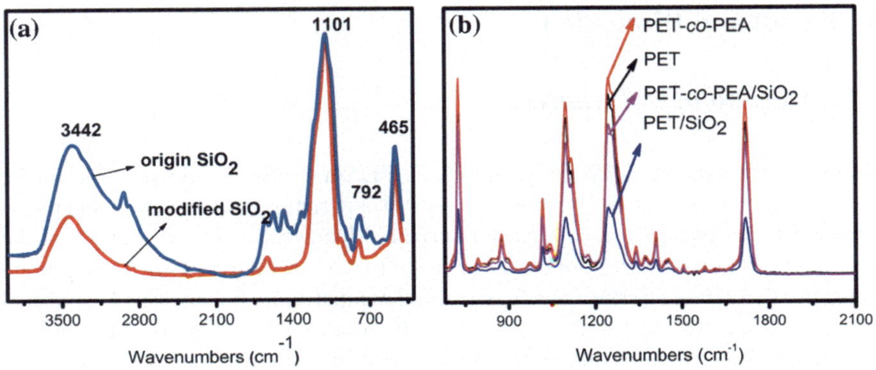

Fig. 4 FTIR spectra of SiO_2 and modified SiO_2 (**a**) and FTIR spectra of PET and its copolyesters (**b**)

Table 2 DSC data of PET copolyesters

Sample	T_g/°C	T_c/°C	T_m/°C	ΔT_c/°C
PET	73.1	137.5	246.1	64.4
PET/SiO_2	72.4	134.1	248.2	61.7
PET-*co*-PEA	63.8	119.6	241.0	55.8
PET-*co*-PEA/SiO_2	61.5	116.6	242.0	55.1

interaction between esters groups in PET and SiO_2 [10]. PET-co-PEA/SiO_2 peaks basically coincide with the peaks of SiO_2 [11], indicating that the addition of PEA and SiO_2 has not changed PET structure.

3.3 DSC Measurement

DSC curve of PET and its copolyesters after the elimination of heat history are shown in Fig. 5 and corresponding data are summarized in Table 2. Glass transition temperature (T_g) of PET-*co*-PEA was reduced from 73.1 to 63.8 °C and melting point (T_m) reduced from 246.1 to 241.0 °C compared with pure PET, respectively. These results are primarily attributed to the flexibility of molecular chains; the copolymerized monomer PEA added in this experiment is a flexible molecular chain segment. The flexibility of PET copolyester macromolecular chain is improved. In addition, it is also seen that the addition of the copolymerized monomer PEA reduces the T_m of the PET, which is also ascribed to the addition of the copolymerized monomer PEA, which leads to the deterioration of the regularity of the macromolecular chains, melting heat ΔH decrease, and random copolymerization damages the regularity of molecular chains, and the melting entropy increases during the melting process. According to the formula of equilibrium melting point:

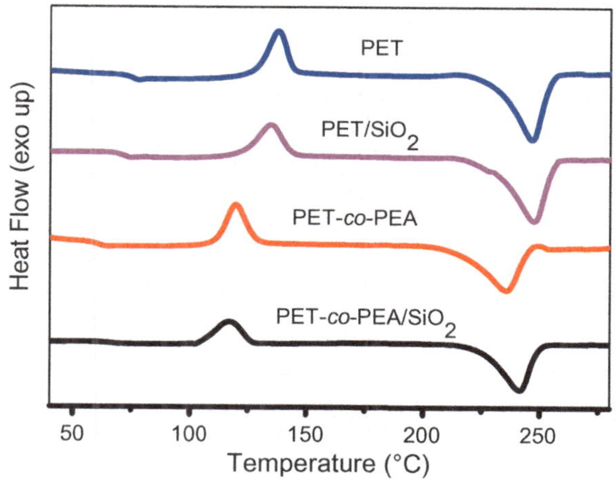

Fig. 5 DSC curves of PET and its copolymers

$$T_{\mathrm{m}}^{0} = \Delta H / \Delta S \quad (1)$$

T_g of PET/SiO_2 has little change compared with neat PET, crystallizing point (T_c) of PET/SiO_2 decreased from 137.5 to 134.1 °C. Degree of super-cooling (T_c) represents capability of the crystallization of copolyester. It can be seen that T_c of PET/SiO_2 is lower than pure PET, which means that it is easy for PET/SiO_2 to crystallize than neat PET. This may attribute to the fact that nano-SiO_2 could act as nucleator to enhance crystallinity of PET copolyesters [12–14]. Furthermore, T_m of PET/SiO_2 increased from 246.1 to 248.2 °C compared with neat PET. However, comparing with neat PET, T_g of PET-*co*-PEA/SiO_2 decreased from 73.1 to 61.5 °C, T_m of PET-*co*-PEA decreased from 242.0 to 246.1 °C, which mainly attributed to synergetic effect of PEA and SiO_2.

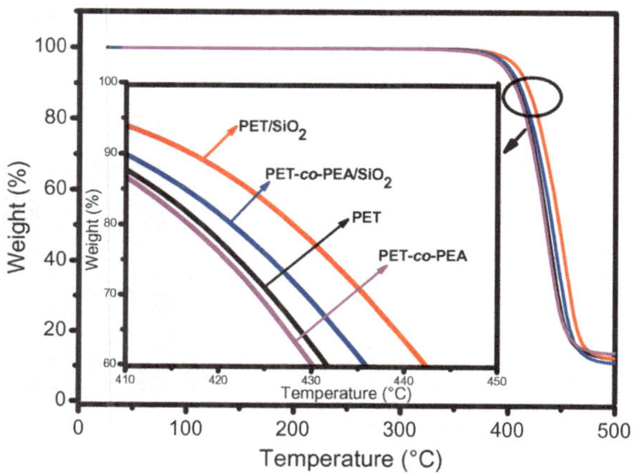

Fig. 6 TGA curves of PET and its copolymers

3.4 TG Analysis

TGA curve of PET and its copolyesters are shown in Fig. 6 and corresponding data are summarized in Table 3. $T_{d(95\%)}$ are temperature with 95% sample mass retention. T_{max} stands for the temperature at fastest thermal degradation rate. Apparently, thermal stability of PET-*co*-PEA has a slight decreased compared with neat PET, which ascribed to thermal stability of PEA are weaker. However, thermal stability of PET/SiO_2 enhanced slightly compared with neat PET, which might be ascribed to strong interfacial interaction between PET chains and modified SiO_2.

3.5 Contact Angle Analysis

Figure 7 shows the values of contact angle measured for PET and its copolyesters. Contact angle of neat PET is 108.59°, contact angle of the PET-co-PEA is 79.05°, contact angle of the PET/SiO_2 is 71.23°, and contact angle of the PET-co-PEA/SiO_2 is 69.56°, respectively. Contact angle of PET gradually decreased with increasing of PEA and SiO_2 amount. This may be due to many strong hydrophilic group and hydroxyl groups (–OH) on the surface of SiO_2. Furthermore, PEA long chains improves compliance of PET, which makes the relative motion of the PET-*co*-PEA

Table 3 TGA data of PET and its copolyesters

Sample	$T_{d(95\%)}$/°C	T_{max}/°C
PET/SiO_2	400.22	438.14
PET	398.89	435.12
PET-*co*-PEA/SiO_2	398.18	434.61
PET-*co*-PEA	393.31	431.34

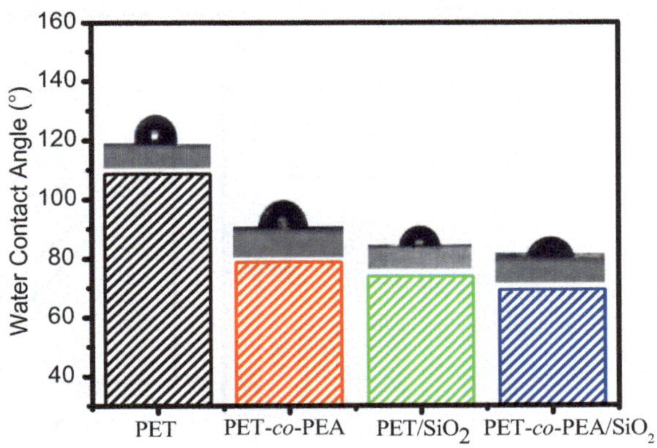

Fig. 7 Water contact angle of PET and its copolymers

molecular chain more easy, the free volume is larger, and the hydrophilicity is more better.

4 Conclusion

In this work, SiO_2 was modified by KH-560, and then samples of PET-*co*-PEA, PET/SiO_2, and PET-*co*-PEA/SiO_2 were prepared. The dispersion of SiO_2 in PET-co-PEA/SiO_2 was tested by field emission scanning electron microscopy (FE-SEM). Subsequently, the crystallization behavior and thermal stability of PET and its copolyesters were measured by differential scanning calorimeter (DSC) and Thermogravimetric analysis (TGA).

(1) FE-SEM and TEM results show that the surface of SiO_2 modified by KH560 was dispersed homogeneously.
(2) FTIR test results show that the addition of PEA and SiO_2 do not change the PET structure. The contact angle test results show that the addition of PEA and SiO_2 can improve PET hydrophilicity.
(3) DSC results show that T_g and T_m of PET-co-PEA/SiO_2 decrease compared with the neat PET. While TG results show that the thermal stability of PET-co-PEA/SiO_2 has little change compared with the neat PET.

References

1. K. Kojima, Y. Takai, M. Ieda, Electroluminescence in polyethylene terephthalate (PET) (impulse voltage), in *Conference on Electrical Insulation & Dielectric Phenomena Report 1981* (2016), pp. 26–31
2. C. Huang, Y.C. Chang, S.Y. Wu, Thin Solid Films **518**(13), 3575–3580 (2010)
3. L. Zhang, W.S. Chin, W. Huang, J.Q. Wang, Surf. Interface Anal. **28**(1), 16–19 (2015)
4. R. Choudhary, A. Kumar, K. Murkute, Periodica Polytech. Civ. Eng. **62**(3) (2018)
5. X. Ren, H.B. Kocer, L. Kou, S.D. Worley, R.M. Broughton, Y.M. Tzou, T.S. Huang, J. Appl. Polym. Sci. **109**(5), 2756–2761 (2010)
6. Y. Wang, C. Xiao, X. Jin, S. An, G. Jia, China Synth. Fiber Ind. (2005)
7. H.Y. Tang, J.Y. Chen, Y.H. Guo, Mater. Des. **31**(7), 3525–3530 (2010)
8. Y. Huang, Y. Zhang, H. Qiu, Y. Ge, Chin. J. Sci. Instrum. **30**(5), 949–953 (2009)
9. X. Cui, B. Zhao, L. Zhang, A. Liu, D. Li, New Chem. Mater. (2015)
10. F.S. Luo, Adv. Mater. Res. **233–235**, 1989–1993 (2011)
11. Y. Wang, L.Y. Wang, J.G. Tang, B. Yang, J. Mater. Eng. **36**(10), 101–105 (2008)
12. J.P. He, H.M. Li, X.Y. Wang, Y. Gao, Eur. Polym. J. **42**(5), 1128–1134 (2006)
13. T.G. Gopakumar, J.A. Lee, M. Kontopoulou, J.S. Parent, Polymer **43**(20), 5483–5491 (2002)
14. S. Tzavalas, V. Drakonakis, DEM, Dieter Fischer A, and VGG. Macromolecules **39**(26), 166–173 (2006)

Preparation and Oxidation Behavior of SiO_2/SiC Coating on Braided Carbon Fiber

Aiming Bu, Yongfu Zhang, Yuping Zhang, Weiwei Chen, Huanwu Cheng and Lu Wang

Abstract People pay more attention to improving the high-temperature oxidation resistance of carbon fiber in aerobic environment. In this study, we proposed a novel method to overcome these critical problems. SiO_2/SiC coating was successfully prepared on braided carbon fiber surface by electrolytic plasma spraying. The scanning electron microscopy (SEM) and X-ray photoelectron spectroscopy (XPS) results showed the changes of surface morphology and surface composition by depositing the coatings, respectively. The energy dispersive spectroscopy (EDS) was employed for identifying the elemental composition of the deposited coating. Thermogravimetric (TG) and differential thermal analysis (DTA) results revealed that the high-temperature oxidation stability of the coated braided carbon fiber was significantly improved.

1 Introduction

Carbon fiber braided has excellent engineering properties, such as a high strength-to-weight ratio, good corrosion and fatigue resistance, and high temperature stability under non-oxidizing atmosphere, thus attracting much attention in a wide variety of fields that require high strength and lightweight properties [1–8]. However, the carbon fiber starts oxidizing at temperatures above 673 K in an oxygen-containing atmosphere leading to the deterioration of the material and thereby degrading the quality of the end application [9, 10]. Coating the carbon fiber braided with ceramic film is an efficient method to enhance the property of the fiber against oxidation [11, 12]. Si-based coatings have excellent oxidation resistance. There are many ways to deposition coating, the methods include physical vapor deposition, plasma treatment, chemical vapor deposition method, solution gelatin coatings, dip coating, and spin coating [13–19]. However, these methods are not an effective approach to preparing

A. Bu · Y. Zhang · Y. Zhang · W. Chen (✉) · H. Cheng · L. Wang
Department of Materials Science and Engineering, Beijing Institute of Technology, Beijing 100081, China
e-mail: wwchen@bit.edu.cn

Y. Han (ed.), *Physics and Techniques of Ceramic and Polymeric Materials*,
Springer Proceedings in Physics 216,
https://doi.org/10.1007/978-981-13-5947-7_18

coatings on the carbon fiber surface. Our previous work developed a novel electrolytic plasma spraying on the carbon fiber bundle.

The electrolytic plasma spraying process is a hybrid of conventional plasma electrolysis and plasma spray. Plasma electrolysis involves electrolysis and electrical discharge phenomena. The surface-treated workpiece is placed at the cathode and the production of an electrical discharge at the workpiece surface. However, the high temperature and strong shock wave generated by the instantaneous arc will have a loss of the substrate, so there is a requirement for the strength of the substrate [20, 21]. Plasma spraying technology uses plasma arc as a heat source which is used to melt materials, such as powder, alloys, ceramics into a molten or semi-molten state by expanding the process gas sprays at a high speed onto the surfaces to create a coating buildup [12]. The limitations of such a process are the severe contact stresses and the erosion of the substrate that are results of the high velocity of the sprayed particles [22, 23]. The electrolytic plasma spraying combines the advantages of plasma electrolysis and plasma spray. The plasma arc is produced by the cathode conducting wire and it forms a plasma stream which by hydraulic pressure and through a plasma nozzle is directed onto the surface to be treated. It has the distinctive advantages such as (1) low capital and operating cost because of relatively simple equipment requirements, (2) producing uniform coating layers for complex components, and (3) low substrate loss and easy to replace cathode. Exhilaratingly, several attempts have been made to fabricate dense, uniform, and oxidation resistance SiO_2/SiC coating on carbon fiber braided using electrolytic plasma spraying method.

This approach opens a new road for preparing coating on carbon fiber braided, and also can be expected to extend to the fabrication of other oxide coatings on complex substrate. Therefore, it is an emerging, environmentally friendly surface engineering technology, and it is of important significance whether in applied or fundamental research fields.

2 Materials and Methods

2.1 The Novel Electrolytic Plasma Spraying

The commercialized carbon fiber braid was used as the substrate. The electrolyte was 30 g/L $Na_2SiO_3{\cdot}9H_2O$ aqueous solution. Figure 1a shows the experimental apparatus of a novel electrolytic plasma spraying process, which copper rod as cathode and graphite as the anode. The surface-treated workpiece is placed under the nozzle. Circulate the electrolyte with circulating pump and discharge the electrolyte from the nozzle. Figure 1b shows the *i-t* diagram for the process of electrolytic plasma spraying and the arcing sequence of the cathodic. When the power is on, hydrogen gas bubbles will be initially produced around the cathode by increasing the current. The quantity of hydrogen bubbles is proportional to the current density. When the hydrogen bubbles reach a certain level, a bubble film wraps around and insulates the

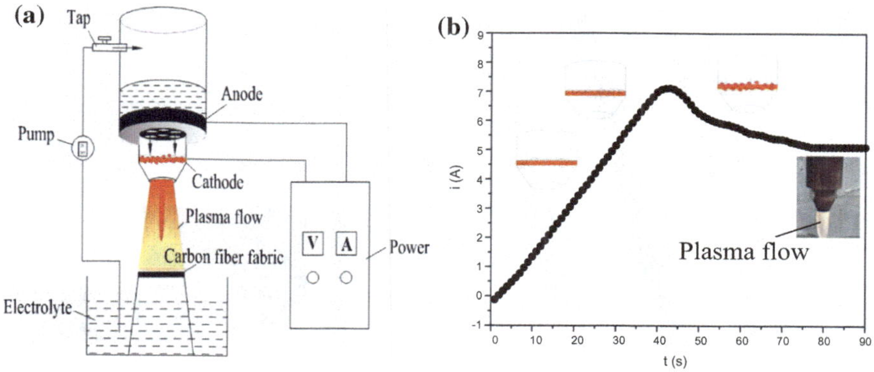

Fig. 1 Schematic diagrams of the electrolytic plasma spraying: **a** the experimental apparatus for depositing coating on carbon fiber fabric by the electrolytic plasma spraying process and **b** the *i-t* diagram for the process of electrolytic plasma spraying and the arcing sequence of the cathodic

cathode. A micro-plasma is discharged across the hydrogen bubble film insulation by increasing DC voltage. The plasma stream is ejected from the nozzle by hydraulic pressure that is directed onto the surface of the carbon fiber braided, as shown in Fig. 1b. We obtained the best carbon fiber braid oxidation resistance by changing the spraying distance and spraying time. The weight-loss rate of coated sample is only 50% when the bare sample has completely lost weight.

2.2 *Characterization Methods*

The microstructures and elemental analysis of the coating were analyzed using a scanning electron microscope (SEM) and energy dispersive spectroscope (EDS). The chemical compositions as well as chemical binding states were characterized using an X-ray photoelectron spectroscopy (XPS). The thermal stability of sample was investigated by thermogravimetry (TG) under air atmosphere, from room temperature up to 1300 °C with a heating rate of 10 °C/min.

3 Results and Discussion

3.1 *Characterization of Coating*

The SEM images and EDS analysis of SiO_2/SiC coating on carbon fiber fabric at deposition time 20 s and spraying distance 15 mm were shown in Fig. 2. Obviously, a dense, uniform, and continuous SiO_2/SiC ceramic coating is distributed on each fiber

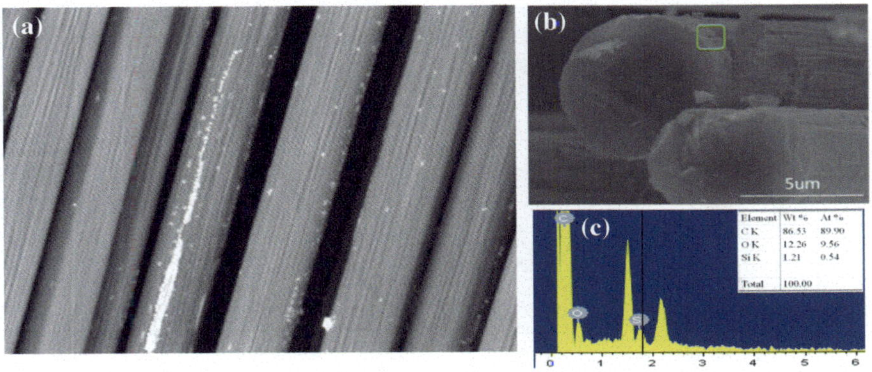

Fig. 2 **a** SEM image, **b** the magnification of the SiO_2/SiC coating on carbon fiber fabric, and **c** EDS analysis at deposition time of 30 s and spraying distance of 20 mm

after deposition of 20 s at room temperature (Fig. 2a). When the distance between the nozzle and the substrate is 15 mm, the plasma is sprayed on the surface of carbon fiber braided with electrolyte, and each fiber was actually affected by the physic-chemical effects to form SiO_2/SiC coating on each fiber. Furthermore, the EDS pattern of the coating was shown in Fig. 2c, indicated that the coating was composed of silicon and carbon. The EDS analysis revealed the Si contents as 1.21 wt%.

3.2 XPS Analysis

The surface chemical composition of carbon fiber fabric and coated sample investigated by XPS was shown in Fig. 3a. The XPS spectrum demonstrates clearly that C1s and O1s are the main surface constituents on both carbon fiber fabric and coated sample surfaces, and a small concentrations of N2p is found. In addition, characteristic Si peak can be observed on coated sample. Figure 3b, c are the XPS of C1s and Si2p spectra were fitted as described, respectively. It can be seen that the C1s peak at 286.1 eV is contributed from C–O and the peak at 284.4 eV is contributed from C–C (Fig. 3b). As shown in Fig. 3c, the Si2p at 101.7 eV is contributed from Si–C and the peak at 102.1 eV is contributed from Si–O.

3.3 Oxidation Resistance

TG under air atmosphere was performed in order to investigate the oxidation stability as well as mass change of the samples. Figure 4 shows the TG and DTA curves of uncoated carbon fiber fabric and coated sample. Figure 4 shows that the mass change slope of coated sample is slower than that of the uncoated sample. Obviously, the

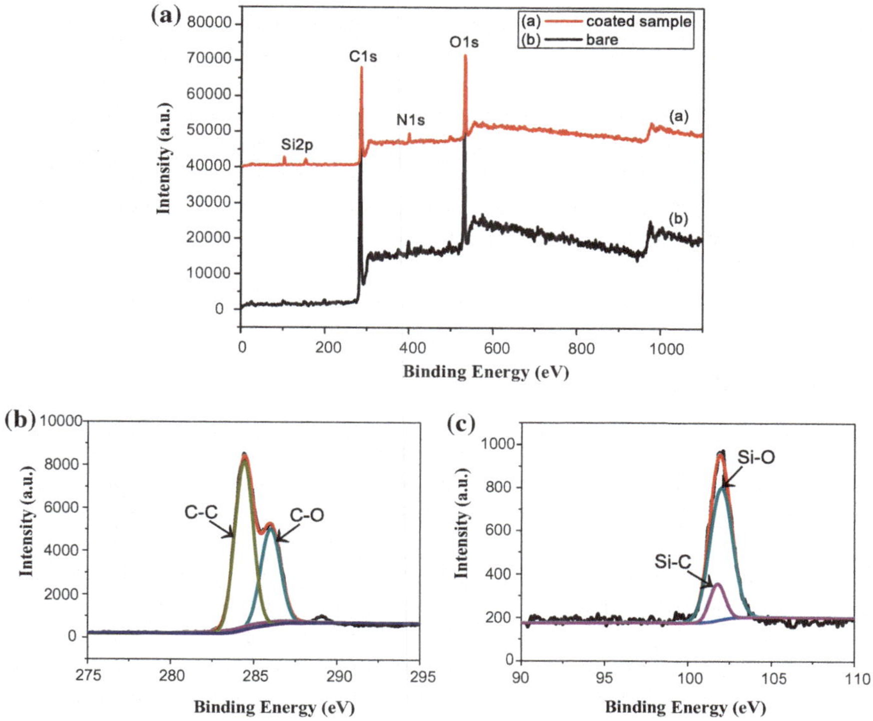

Fig. 3 XPS spectra of **a** uncoated sample and coated sample and **b** C1s and **c** Si2p peaks from XPS spectra

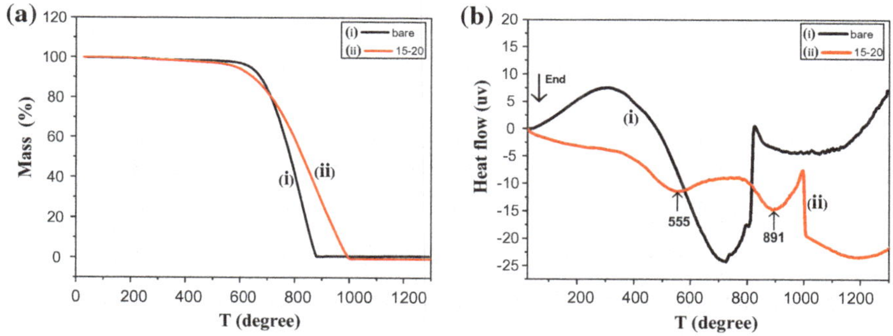

Fig. 4 **a** TG curves and **b** DTA curves of the uncoated and coated samples at deposition time of 20 s and spraying distance of 15 mm

curves are almost horizontal for both types of fibers at an oxidation temperature of 600 °C. With the extension of oxidation time and the increase of oxidation temperature, the weight loss is also starting to increase. The mass losses of uncoated carbon fiber fabric sample are found to be 59 and 99% at 800 and 900 °C, respectively. However, the mass loss is found to be 41 and 73% for coated sample at the same oxidation conditions. When the temperature rises to 1000 °C, the coated sample losses weight completely, which states an excellent oxidation resistance. Two typical endothermic peaks were observed at 555 and 891 °C for the SiO_2/SiC-coated carbon fiber (Fig. 4b).

4 Conclusions

In this paper, we designed the experimental devices by combining plasma electrolysis and plasma spraying. A dense, uniform, and continuous SiO_2/SiC coating was successfully prepared on the carbon fiber fabric surface by electrolytic plasma spraying. Sample preparation was adjusted by changing the spray distance and spray time. The oxidation resistance of coatings was studied by thermogravimetric method. The results show that the bare carbon fiber fabric suffered from burnout at 900 °C. However, when the spraying distance is 15 mm and the spraying time is 20 s, the coated sample burned out at 1000 °C. The outstanding oxidation resistance coating was deposited on carbon fiber fabric by this new method and device. It shows a promising use in the composite materials.

Acknowledgements We thank Dr. Maoyuan Li and Dr. Lin Lu for helpful discussions.

References

1. M. Ghanbarian, E.T. Nassaj, A. Kariminejad, Synthesis of nanostructural turbostratic and hexagonal boron nitride coatings on carbon fiber cloths by dip-coating. Surf. Coat. Technol. **288**, 185–195 (2016)
2. K. Kim, Y.C. Jung, S.Y. Kim, B.J. Yang, J. Kim, Adhesion enhancement and damage protection for carbon fiber-reinforced polymer (CFRP) composites via silica particle coating. Compos. Part A **109**, 105–114 (2018)
3. Y.Q. Wang, B.L. Zhou, Z.M. Wang, Oxidation protection of carbon-fibers by coatings. Carbon **33**, 427–433 (1995)
4. K.D. Xia, C.X. Lu, Y. Yang, Preparation of anti-oxidative SiC/SiO_2 coating on carbon fibers from vinytriethoxysilane by sol-gel method. Appl. Surf. Sci. **265**, 603–609(2013)
5. G. Wu, L. Ma, L. Liu, Y. Wang, F. Xie, Z. Zhong, M. Zhao, B. Jiang, Y. Huang, Interface enhancement of carbon fiber reinforced methylphenylsilicone resin composites modified with silanized carbon nanotubes. Mater. Des. **89**, 1343–1349 (2016)
6. J.J. Wang, W.S. Lin, X. Wu, Y.Y. Yang, F. Wu, X.Z. Yan, Preparation and oxidation behavior of a double-layer coating for three-dimensional braided carbon fiber. Surf. Coat. Technol. **298**, 58–63 (2016)

7. S.F. Tang, C.L. Hu, Design, preparation and properties of carbon fiber reinforced ultra-high temperature ceramic composites for aerospace applications: a review. J. Mater. Sci. Technol. **33**, 117–130 (2017)
8. R. Naslain, Design, preparation and properties of non-oxide CMCs for application in engines and nuclear reactors: an overview. Compos. Sci. Technol. **64**, 155–170 (2004)
9. M. Niu, H. Wang, J. Wen et al., Preparation and anti-oxidation properties of Si(O)C coated carbon-bonded carbon fiber composites. RSC Adv **5**(65), 52347–52354 (2015)
10. B. Du, C.Q. Hong, X.H. Zhang, A.Z. Wang, Y.Q. Sun, Ablation behavior of advanced $TaSi_2$-based coating on carbon-bonded carbon fiber composite/ceramic insulation tile in plasma wind tunnel. Ceram. Int. **44**, 3505–3510 (2018)
11. F. Alonso, I. Fagoaga, P. Oregui, Erosion protection of carbon—epoxy composites by plasma-sprayed coatings. Surf. Coat. Technol. **49**, 482–488 (1991)
12. A. Lopera-Valle, A. McDonald, Application of flame-sprayed coatings as heating elements for polymer-based composite structures. J. Therm. Spray Technol. **24**, 1289–1301 (2015)
13. J.F. Ma, W.F. Fu, Y.Q. Meng, Z.Q. Yu, S. Cai, B.B. Niu, "Electrochemical" growth of ZnO coating on carbon fiber. Mater. Chem. Phys. **171**, 22–26 (2016)
14. R. Pillai, N. Batra, L.M. Manocha, N. Machinewala, Deposition of silicon carbide interface coating on carbon fibre by PECVD for advanced composites. Surf. Interfaces **7**, 113–115 (2017)
15. M. Wang, X.G. Diao, A.P. Huang, P.K. Chu, Z. Wu, Influence of substrate bias on the composition of SiC thin films fabricated by PECVD and underlying mechanism. Surf. Coat. Technol. **201**, 6777–6780 (2007)
16. P. Bertrand, M. Vidal-Setif, R. Mevrel, LPCVD pyrocarbon coating on unidirectional carbon fiber yarns: an efficient interphase for aluminium matrix composites. J. Phys. **5**(5), 769–776 (1995)
17. H.J. Yuan, C.X. Lu, S.C. Zhang, G.P. Wu, Preparation and characterization of a polyimide coating on the surface of carbon fibers. New. Carbon. Mater. **30**(2), 115–121(2015)
18. A. Kariminejad, E. Taheri-Nassaj, M. Ghanbarian, S.A. Hassanzadeh-Tabrizi, Effects of PACVD parameters including pulsed direct current and deposition time on nanostructured carbon coating deposited on carbon fiber fabrics. Mater. Des. **106**, 184–194 (2016)
19. M.A. Montes-Morán, F.W.J.V. Hattum, J.P. Nunes, A. Martínez-Alonso, J.M.D. Tascóna, C.A. Bernardo, A study of the effect of plasma treatment on the interfacial properties of carbon fibre-thermoplastic composites. Carbon **43**, 1795–1799 (2005)
20. E.I. Meletis, X. Nie, F.L. Wang, J.C. Jiang, Electrolytic plasma processing for cleaning and metal-coating of steel surfaces. Surf. Coat. Technol. **150**, 246–256 (2002)
21. A.D. Lin, C.M. Hsu, C.K. Chen, D.Y. Huang, T.P. Hung, J.H. Kuang, Stainless steel surface etching morphologies using electro plasma technology. Adv. Sci. Lett. **13**, 178–182 (2012)
22. R. Lupoi, W. O'Neill, Deposition of metallic coatings on polymer surfaces using cold spray. Surf. Coat. Technol. **205**, 2167–2173 (2010)
23. A. Rezzoug, S. Abdi, A. Kaci, M. Yandouzi, Thermal spray metallisation of carbon fibre reinforced polymer composites: effect of top surface modification on coating adhesion and mechanical properties. Surf. Coat. Technol. **333**, 13–23 (2018)

Enhanced Biocompatibility via Adjusting the Soft-to-Hard Segment Ratios of Poly (Ether-Block-Amide) Medical Hollow Fiber Tube for Invasive Medical Devices

Z. M. Li, Y. Y. Xue, Z. H. Tang, S. Zhu, M. L. Qin and M. H. Yu

Abstract Poly(ether-block-amide) (Pebax) is a promising polymeric material for the application in biomedical area, which is significantly influenced by its property of biocompatibility. For segment copolymers like Pebax-family materials, the biocompatibility is closely related to the soft-to-hard segment ratios. This study aims to explore the relationship between the biocompatibility property and soft-to-hard segment ratios of Pebax medical hollow fiber tube for invasive medical devices. Various analytical techniques including Fourier transform infrared spectroscopy (FTIR), X-ray diffraction (XRD), atomic force microscope (AFM), scanning electron microscopy (SEM) and dynamic thermomechanical analysis (DMA) were used to observe the Pebax hollow fiber tube with various soft-to-hard segment ratios. The results indicated that the degree of microphase separation increased with the decrease of the soft segment ratios of Pebax hollow fiber tube. Similarly, we have clearly observed the effect of soft-to-hard segment ratios in the wettability of Pebax films with the contact angle method. The biocompatibility of Pebax hollow fiber tube was characterized by hemolysis tests and vitro cytotoxicity tests. It was found that the biocompatibility was greatly influenced by the content of soft segment ratios. The satisfactory biocompatibility property can be achieved via reducing the soft segment ratios of Pebax hollow fiber tube.

Z. M. Li · S. Zhu · M. L. Qin · M. H. Yu (✉)
State Key Laboratory for Modification of Chemical Fibers and Polymer Materials, College of Materials Science and Engineering, Donghua University, Shanghai 201620, People's Republic of China
e-mail: Yumuhuo@dhu.edu.cn

M. L. Qin
e-mail: mlqin@microport.com

Z. M. Li · M. L. Qin
AccuPath Medical (Jiaxing) Co. Ltd., Shanghai, China

Y. Y. Xue · Z. H. Tang
School of Materials Science and Engineering, University of Shanghai for Science and Technology, Shanghai 200433, China

Y. Han (ed.), *Physics and Techniques of Ceramic and Polymeric Materials*, Springer Proceedings in Physics 216,
https://doi.org/10.1007/978-981-13-5947-7_19

1 Introduction

Poly(ether-block-amide) (Pebax) is a category of multiblock copolymers consisting of polyamide (PA) as hard segment and the flexible soft segment polyether (PE) [1]. Pebax has attracted considerable attention in the biomedical field due to its excellent properties, such as good biodegradability, shape memory property and biocompatibility. Moreover, Pebax exhibits a wide range of mechanical behaviors from thermoplastics to elastomers which depends on its soft segment components [8]. The thermodynamic incompatibility between soft and hard segments of Pebax was found to result in microphase separation with diverse morphologies [2–6].

Pebax can be processed to make small-diameter medical catheters, which can be used to treat diseased blood vessels, clean out blocked vessels, and deliver drugs and clot-dissolving medications directly to the target area [7]. Nowadays, Pebax has been applied in many fields such as seals, automotive clip gaskets, air bag covers, pneumatic hoses [1, 4], angiography, angioplasty and urology catheters [7, 8], catheter sheaths and catheter tips [9].

Owing to different soft segments of Pebax with different properties, numerous studies have been conducted to explore the relationship of soft segment ratios of Pebax to that of its properties and microphase separation. For example, Wilkes and coworkers found that microphase separation was related to the soft segment ratios of Pebax by investigating the solid state structure-property behavior of Pebax with different soft segment contents [3]. Gilchrist et al. studied the effect of extreme processing on morphology and mechanical properties of Pebax in microinjection molding process [10]. However, they only focused on the effect of chemical composition of Pebax on microphase separation morphology and related properties. To the best of our knowledge, there are still not any published reports to investigate the effect of the various soft-to-hard segment ratios of Pebax hollow fiber tube on the biocompatibility property. Therefore, it seems very meaningful to explore the relationship between the various soft-to-hard segment ratios of Pebax hollow fiber tube and the biocompatibility property, which may conduce to screen the most suitable biomaterials with good biocompatibility for biomedical application.

The present study aims to investigate the relationship between the various soft-to-hard segment ratios of Pebax hollow fiber tube and the biocompatibility property in detail. Pebax hollow fiber tube with different soft segments was characterized by various analytical techniques, such as Fourier transform infrared spectroscopy (FTIR), X-ray diffractometer (XRD), scanning electron microscopy (SEM), atomic force microscope (AFM) and Dynamic thermomechanical analysis (DMA). The biocompatibility of Pebax hollow fiber tube was evaluated by the hemolysis and cytotoxicity test. The results indicated that a decrease in the soft segment components tends to increase the degree of microphase separation and the biocompatibility of Pebax hollow fiber tube.

2 Experiment

2.1 *Materials*

Pebax (XX33) grades were supplied by Arkema in pellet form, polyamide 12 (PA 12) as the hard segment and polytetramethylene oxide (PTMO) as the soft segment. XX denotes the shore D hardness, such as 25, 35, 40, 63, 70, and 72. All resins were dried at 90 °C in an oven for overnight before use.

2.2 *Preparation of Pebax Hollow Fiber Tube with Various Soft-to-Hard Segment Ratios*

Dried Pebax particles with varying soft-to-hard segment ratios were melted and extruded into fiber tube by single-screw precision extruder (American Kuhne, USA) at 200–210 °C. The melt-extruded fiber tube was drawn with a drawing rate of 3 m/min in water. The temperature of water coagulation bath was 25 °C.

2.3 *Preparation of Pebax Film with Various Soft-to-Hard Segment Ratios*

Pebax particles with varying soft-to-hard segment ratios were dried in a vacuum oven at 90 °C for overnight and then poured into the mold, the samples were melted at 210 °C for 5 min under 1.5 MPa pressure. Then, the specimens were slowly cooled to room temperature.

2.4 *Characterization*

FTIR spectra (Nicolet 8700) were obtained by using a Nicolet FT-IR spectrometer with a Perkin-Elmer Frustrated Multiple Internal Reflections 186-0174 accessory. The spectra of the samples were collected with a resolution of 4 cm^{-1}. The FTIR data were collected between 600 and 4000 cm^{-1}. The crystalline of the Pebax hollow fiber tube with various soft-to-hard segment ratios was monitored with an X-ray diffractometer (D/MAX-2500/PC, Rigaku Corporation) using Cu–Ka radiation at a generator voltage of 40 kV and current of 40 mA. The data were obtained from $2\theta =$ 10–60° at a scanning speed of 2°/min. The fracture morphologies of Pebax hollow fiber tube with soft-to-hard segment ratios were observed with an FE-SEM (JEOL JSM-6330F), all samples were coated with a layer of gold or platinum. The samples

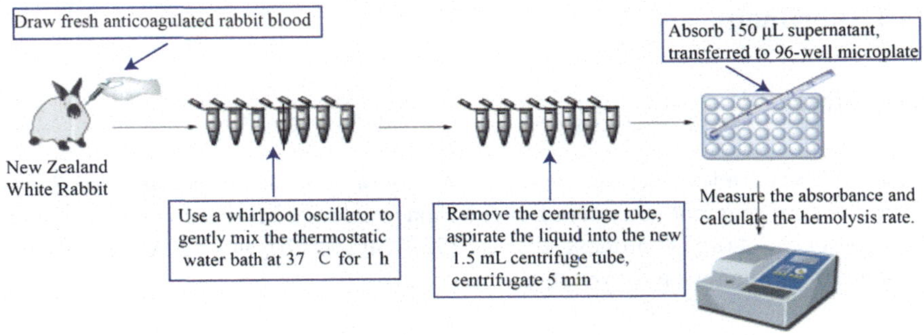

Fig. 1 The procedure of hemolysis test

were placed onto mica plate and the microcosmic surface shape was observed by the AFM (American Veeco, IIIa MultiMode) with the scope size of 2 μm (scan rate 0.5 Hz). The wettability of Pebax films was measured by the contact angle of the water according to the sessile drop technique implemented on a Dataphysics OCA 40 contact angle system. For contact angle measurements, the Pebax films were fixed on a frame, and the volume of the water droplets was 5 μL. DMA measurements were performed via a DMA Q800 (TA Instruments). The samples were heated under a dry nitrogen atmosphere from −150 to 150 °C at the rate of 2 °C/min; Tg data were obtained at a frequency of 1 Hz. Pebax hollow fiber tube was sealed before packaging and then analyzed at low temperature for 36 h in ethylene oxide. The hemolysis tests were conducted according to the standard GB/T16886 12-2005. Fresh anticoagulation rabbit blood was obtained from New Zealand white rabbits. Figure 1 shows the major procedure of hemolysis tests. The absorbance was measured at 545 nm of the microplate reader and the hemolysis rate (HR) were calculated by the following formula:

$$\mathrm{HR} = (A - B)/(C - B) \times 100\%$$

where: A is the test group absorbance, B is the negative control group absorbance, C is the positive control group absorbance. The test group and the control group absorbance to take the average of three tubes.

With regard to in vitro cytotoxicity test, L929 cells (Mouse fibroblasts, ATCC CCL1) were grown in a 25 cm^2 culture flask and grown to ≥80%. Fetal bovine serum MEMα medium (10%, without any sample), high-density polyethylene (HDPE), and DMSO solution (5%, dimethylsulfoxide solution) were used as the blank control, negative control and positive control, respectively. Each group was pre-cultured at 37 °C and 5% CO_2 for 72 h, and the change of cell status in each column was observed every 24 h under a microscope. For three-point bending experiments, Pebax hollow fiber tube with various soft segment components (each length requires 100 ± 1 mm) was tested and kept balance for 48 h on the constant temperature and humidity box (25 °C, 60%). The test speed was 5 mm/min, and bending displacement was 4 mm.

Ten N sensor was selected for the Instron Universal Tensioner (FM-02-003) (replace the 500 N sensor if the upper limit is greater than 10 N). More than five parallel samples were measured for each test. The resistance to internal pressure of Pebax hollow fiber tube (ID: 0.75 mm, OD: 0.9 mm) was measured by PT-1000.

3 Result and Discussion

3.1 The Structure of Pebax Hollow Fiber Tube with Various Soft-to-Hard Segment Ratios

Figure 2 shows the FTIR spectra of the Pebax hollow fiber tube with various soft-to-hard segment ratios. The absorption peak at 3310 cm^{-1} was assigned to the stretching vibration of hydrogen-bonded N–H in amide groups. It can be found that the absorption intensity of NH increases with the decreasing of Pebax soft segment content, which represents the higher degree of microphase separation. Furthermore, the infrared bands at 1110 cm^{-1} corresponding to the hydrogen-bonded of –C–O–C– groups was observed. The hydrogen bond of –C–O–C– decreases with the decreasing of Pebax soft segment ratios, which also demonstrates that the microphase separation degree of Pebax increases with a decrease in the soft segment ratios of Pebax. In addition, it can be found that the PTMO blocks exhibit an ether group at 1100 cm^{-1}, while the PA 12 block exhibits carbonyl group at 1640 cm^{-1} [11–13].

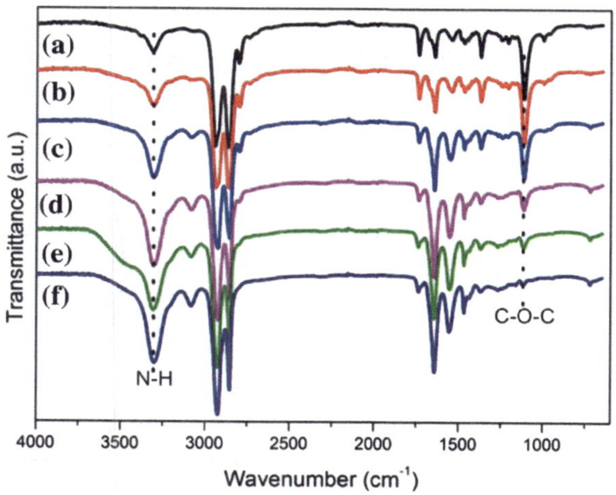

Fig. 2 FT-IR spectra of Pebax hollow fiber tube with various soft-to-hard segment ratios **a**–**f** 2533, 3533, 4033, 6333, 7033 and 7233

3.2 The Crystallinity of Pebax Hollow Fiber Tube with Various Soft Segment Components

The crystalline structures of Pebax hollow fiber tube with various soft-to-hard segment ratios were investigated by XRD analysis. Figure 3 exhibits the XRD patterns of Pebax hollow fiber tube with various soft segment ratios, respectively. It can be seen that only one diffraction at $2\theta = 20.8°$ was observed from Fig. 3a indicating an almost amorphous packing of macromolecules. It might be attributed to an extremely high content of Pebax soft segment, which could lead to the incomplete separation of the soft and hard segment structures and ultimately result in crystallization of the hard segment. As can be seen, a new crystal peak appeared at $2\theta = 21.2°$ and the intensity increased with decreasing the soft segment ratios of Pebax (Fig. 3b–f). This is primarily due to the immobilization of molecular chain rearrangement movement under higher content of Pebax hard segment, and the hard segment was released from the crystalline state of inhibition, which led to more perfect microphase separation. On the other hand, it also indicates that the hard segment content of Pebax leads to increment of their crystallinity. In addition, it can be found that a new crystal peak appeared at $2\theta = 37.3°$ when decreasing the soft segment ratios of Pebax, due to the different crystalline conformation of PA.

3.3 Morphologies of Pebax Hollow Fiber Tube with Various Soft-to-Hard Segment Ratios

The fracture morphologies of Pebax hollow fiber tube with various soft-to-hard segment ratios were investigated by SEM. As shown in Fig. 4a, the morphology of the Pebax (2533, 3533) hollow fiber tube was relatively rough. This is probably due to the

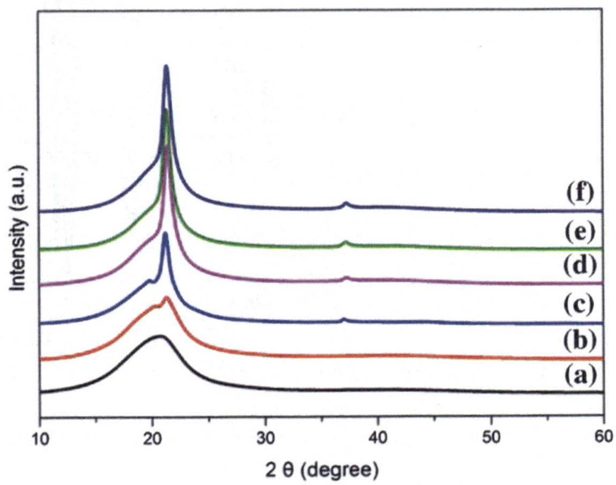

Fig. 3 The crystallinity of Pebax with various soft-to-hard segment ratios **a–f** 2533, 3533, 4033, 6333, 7033 and 7233

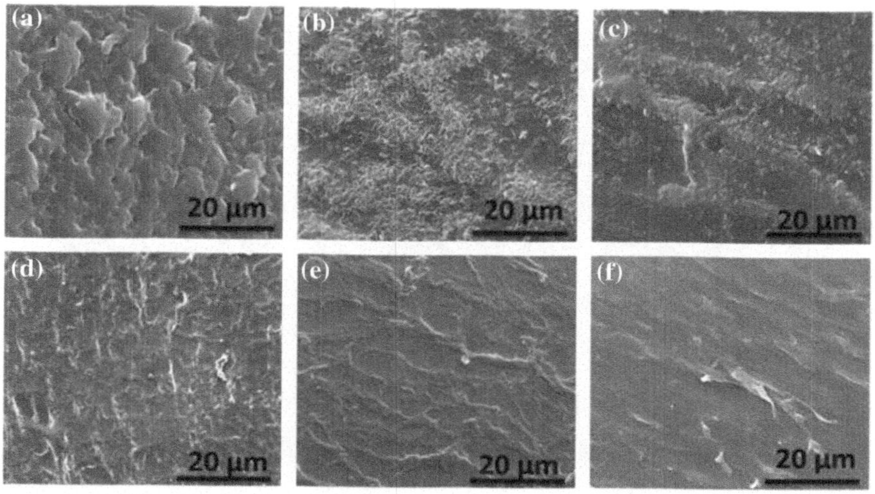

Fig. 4 Morphologies of Pebax hollow fiber tubes with various soft segment components **a–f** 2533, 3533, 4033, 6333, 7033 and 7233

lower degree of microphase separation, which results in the most of soft domains on the polymer surface. However, from Fig. 4b–f, it can be seen that the surface of Pebax with various soft segment ratios became more and more rough. It can be explained that the hard blocks aggregating into the semi-crystalline structure were dispersed into the continuous phase of the soft blocks, which leads to physical cross-linking points and prevents the macromolecular chains from slipping at ambient temperature. Due to the lower surface energy and Tg, soft-phase layer rearranged to cover the surface of PTMO, resulting in the formation of a rough surface with decreased surface area. Meanwhile, due to the low content of hard segment (PA 12), the smaller semi-crystalline particles are dispersed into the soft segment continuous phase, which results in a large number of soft domains on the polymer surface. Besides, as the hard segment increases (PA 12), the larger crystalline particles cause the continuous phase of the PTMO to fail directly through the hard segment region. When the hydrogen bonding interaction in the soft segment becomes weaker, the degree of microphase separation becomes larger between soft and hard segment. Thus, the morphology of Pebax hollow fiber tube becomes smoother.

3.4 Phase Images of Pebax Hollow Fiber Tube with Various Soft-to-Hard Segment Ratios

To further investigate the microphase separation of Pebax hollow fiber tube with soft-to-hard segment ratios, the phase images of Pebax hollow fiber tube were measured by AFM and shown in Fig. 5. As can be clearly seen, the harder phases were rarely

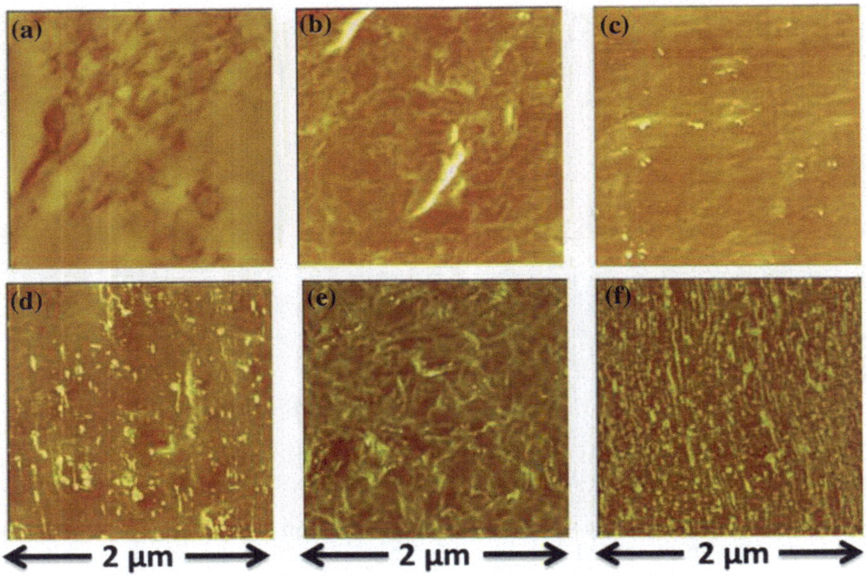

Fig. 5 Phase images of Pebax hollow fiber tubes with various soft-to-hard segment ratios **a**–**f** 2533, 3533, 4033, 6333, 7033 and 7233

observed in Pebax 2533 and Pebax 3533, while the harder phases of Pebax 4033 and 6333 interspersing among softer phases were observed (Fig. 5c, d). Pebax 7033 and 7233 showed higher proportion of harder phases and better dispersion (Fig. 5e, f). The results showed that the microphase separation occurring obviously when the soft segment ratios decreased.

3.5 Wettability of Pebax Films with Various Soft-to-Hard Segment Ratios

The surface hydrophilicity is one of the important factors affecting the biocompatibility of the material [14]. Cell adhesion is mainly determined by surface wettability but is also affected by the surface functional group, its surface density, and the kinds of cells. HUVECs were cultured on the mixed SAMs preadsorbed with albumin. Cell adhesion was effectively prohibited on hydrophobic SAMs pretreated with albumin. Albumin strongly adsorbed and resisted replacement by cell adhesive proteins on hydrophobic SAMs. On the other hand, cells adhered to albumin-adsorbed hydrophilic SAMs. Displacement of preadsorbed albumin with cell adhesive proteins effectively occurs on these hydrophilic SAMs [15]. This effect contributes to induce SAMs with moderate wettability to give suitable surfaces for cell adhesion. The introduction of hydrophobic macromolecules into the surface of the materials provides an

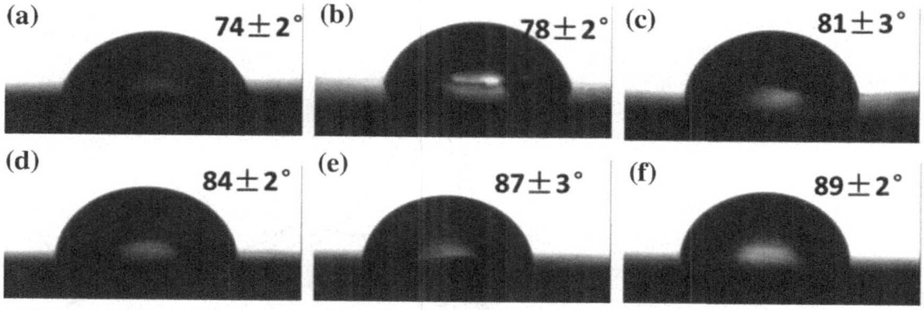

Fig. 6 Static contact angle tests of Pebax films with various soft segment components **a**–**f** 2533, 3533, 4033, 6333, 7033 and 7233

effective way to increase the surface hydrophobicity and reduce the adhesion [16]. In this study, the hydrophilicity of Pebax films with various soft segment components was evaluated by the contact angle test, and the results were shown in Fig. 6. It can be seen that the contact angles of the Pebax films increase with a decrease in the soft segment components, which results in decreased hydrophilicity. The decreased hydrophilicity of Pebax film would lead to the increased biocompatibility as the soft segment components of Pebax decreases.

3.6 The Dynamical Thermal Mechanical Property

In terms of thermoplastic elastomers, the nature of the segments and the block length are the main factors inducing phase segregation. In this study, DMA was used to investigate the microphase separation of Pebax hollow fiber tube with various soft segment components. The loss tangent (tan δ) of Pebax with various soft segment components was plotted in Fig. 7. It can be found that the Tg, s and Tg, h values of Pebax films with various soft segment components based on the peak temperature of tan δ were in the range of −57.5 to −75.5 °C and −9.1 to 34.4 °C. In addition, the Tg, h of Pebax hollow fiber tube increased gradually with the decreasing soft segment components, implying the degree of microphase separation increases. The results in this study were consistent with the previous report [17].

3.7 Biocompatibility of Pebax Hollow Fiber Tube with Various Soft-to-Hard Segment Ratios

Figure 8 shows the hemolysis rate of Pebax hollow fiber tube with various soft-to-hard segment ratios. It can be obviously observed that the hemolysis rate of Pebax increased with increasing soft segment ratios. The results indicated the blood com-

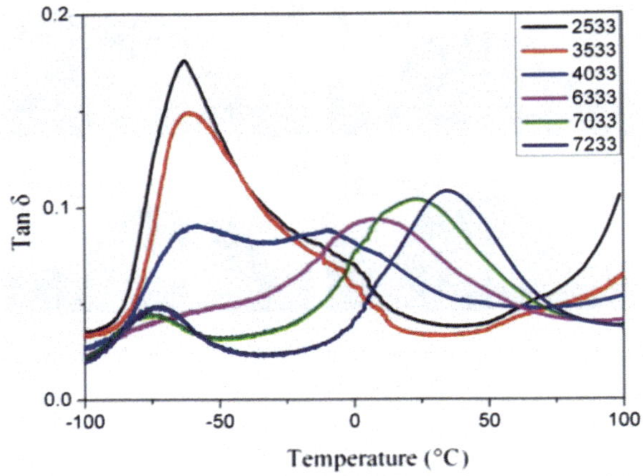

Fig. 7 DMA properties of Pebax hollow fiber tubes with various soft segment components

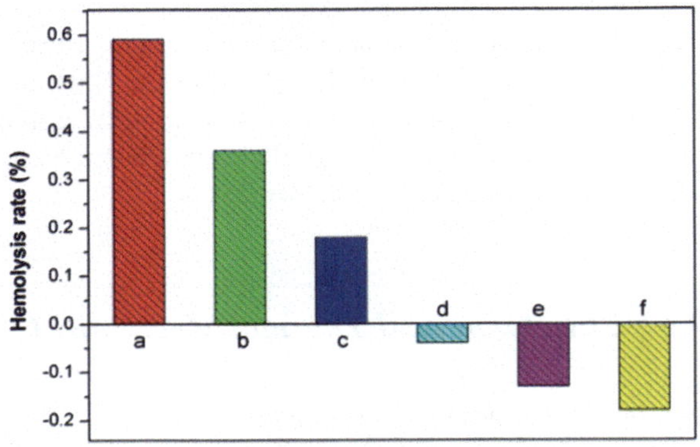

Fig. 8 Hemolysis rates of Pebax hollow fiber tubes with various soft-to-hard segment ratios **a**–**f** 2533, 3533, 4033, 6333, 7033 and 7233

patibility of Pebax improved with decreasing the soft segment ratios. This could be explained by the increased hydrophobicity and blood compatibility of Pebax with the soft segment ratios decreasing, the result was consistent with the contact angle test. In addition, the morphological structure also affects the blood compatibility of polymer, and the higher degree of microphase separation enhances the blood compatibility of polymer.

Figure 9 exhibits the cell growth situation of Pebax hollow fiber tube with various soft-to-hard segment ratios at 72 h. It can be obviously found that the number of cell increased with decreasing the soft segment ratios. This indicates that the microphase separation degree of Pebax increases with an increase in hardness, resulting in better biocompatibility of the polymer with the cells. For one thing, with the increase of hardness, the microstructure of Pebax becomes close to that of the adsorbed proteins, which plays an important role in promoting cell adhesion [18]. For another thing, the surface of Pebax may affect the interaction between the material and the organism, thus affecting the biocompatibility of the material [19]. Under different hardness, the surface of the material can achieve different hydrophobic transition. Low hardness

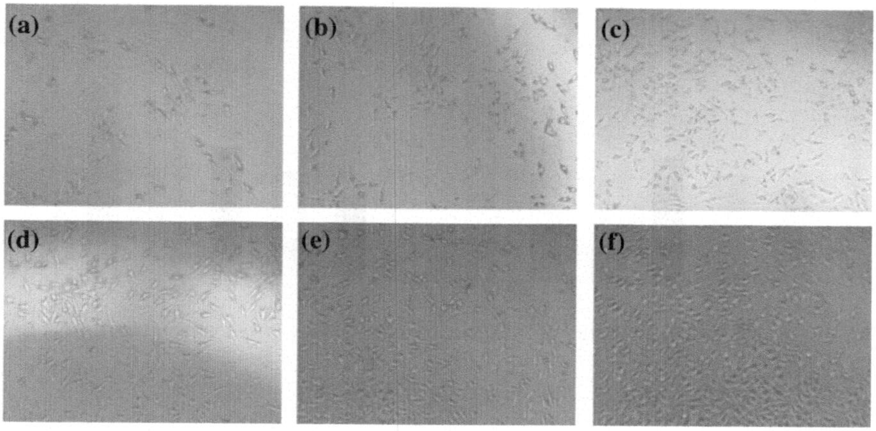

Fig. 9 The cell growth situation of Pebax with various soft-to-hard segment ratios **a–f** 2533, 3533, 4033, 6333, 7033 and 7233

of Pebax causes weak contact reaction of cells and Pebax particles, leading to few ultimate cell numbers. Nevertheless, with the hard content increasing, the content of PA-12 increases with high-contact reaction of cells and Pebax particles. Besides, the increase of crystallinity of the hard segment can lead to a decrease in the compatibility between the hard segment and the soft segment [3]. In addition, the increased degree of microphase separation between the hard segment and the soft segment could lead to the regular surface of Pebax, as determined by SEM and XRD. Regular Pebax surface may benefit the cell contact reaction, which can ultimately increase the cell adhesion and proliferation onto the material surfaces.

3.8 *Mechanical Properties of Pebax Hollow Fiber Tube with Various Soft-to-Hard Segment Ratios*

To further investigate the effect of microphase separation on the mechanical properties of Pebax medical hollow fiber tube with different soft-to-hard segment ratios, the flexural properties and resistance to internal pressure of hollow fiber tube were discussed. As shown in Fig. 10a, it is apparently found that the flexural stress of Pebax hollow fiber was improved significantly with the increase of hard segment ratios. The resistance to internal pressure of Pebax hollow fiber tube was also used to assess the effect of microphase separation on the mechanical properties of Pebax hollow fiber tube with different hard segment ratios. Figure 10b indicated that resistance to internal pressure of Pebax hollow fiber tube was also enhanced significantly with the increase of hard segment ratios. The results indicated that microphase separation had a great impact on the mechanical properties of Pebax hollow fiber tube.

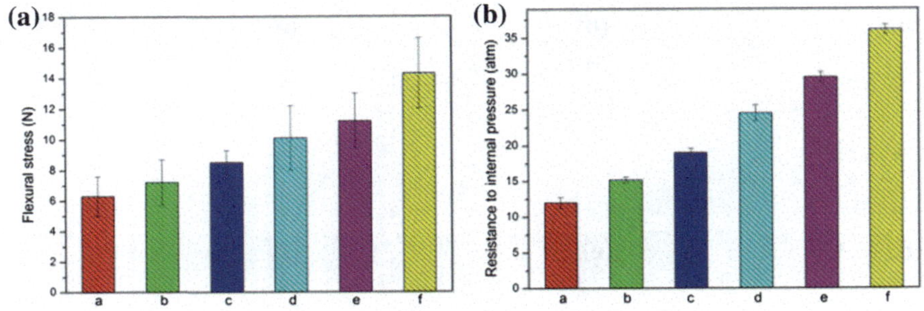

Fig. 10 The mechanical properties of Pebax hollow fiber tubes with various soft segment components

4 Conclusion

In this paper, we systematically investigated the effect of the structure of various soft-to-hard segment ratios of Poly(ether-block-amide) (Pebax) hollow fiber tube on biocompatibility. The results of FT-IR, XRD and SEM showed that the microphase separation degree of Pebax increased with the decrease of Pebax soft segment ratios. Meanwhile, to further investigate the microphase separation of Pebax with various soft-to-hard segment ratios, the wettability and the phase images of Pebax films were measured by static water contact angle test and AFM. The biocompatibility of Pebax with various soft-to-hard segment ratios was enhanced significantly by the higher degree of microphase separation of Pebax. This may be advantageous for the clinicians and practitioners to develop suitable biomaterials for practical application. Furthermore, the mechanical properties of Pebax hollow fiber tube were also investigated, which was found to be significantly affected by the microphase separation.

Acknowledgements This work was financially supported by the National Basic Research Program (2011 CB606101) of the China 973 Program, National Natural Science Foundation of China (No. 21404023), funding from State Key Laboratory for Modification of Chemical Fibers and Polymer Materials, Donghua University (No. LK1515).

References

1. M.C. Choi, J.Y. Jung, H.S. Yeom, Y.W. Chang, Polym. Eng. Sci. **53**, 982 (2013)
2. M.R. Barzegari, N. Hossieny, D. Jahani, C.B. Park, Polymer **114**, 15 (2017)
3. J.P. Sheth, J. Xu, G.L. Wilkes, Polymer **46**, 743 (2003)
4. X.J. Guo, J. Polym. Sci. Part B. Polym. Phys. **46,** 2035 (2008)
5. E.V. Konyukhova, A.I. Buzin, Y.K. Godovsky, Thermo. Acta. **391**, 271 (2002)
6. P. Rangarajan, R.A. Register, D.H. Adamson, L.J. Fetters, W. Bars, S. Naylor, A.J. Ryan, Macromolecules **28**, 1422 (1995)
7. Y. Song, H. Yamamoto, N. Nemoto, Macromolecules **37**, 6219 (2004)

8. C.X. Wang, M. Du, J.C. Lv, Q.Q. Zhou, Y. Ren, G.L. Liu, D.W. Gao, L.M. Jin, Appl. Surf. Sci. **349**, 333 (2015). B. Tavernier, J. Mewis, P.V. Puyvelde, M. Takenaka, B. Ernst, T. Hashimoto, Polym. Eng. Sci. **48**, 2418 (2008)
9. D.A. Mair, E. Geiger, A.P. Pisano, J.M. Frechet, F. Svec, Lab Chip **6**, 1346 (2006)
10. N. Zhang, S.Y. Choi, M.D. Gilchrist, Mater. Eng. **299**, 1362 (2015)
11. Pebax Polyether Block Amides: Product Properties, http://www.Pebax.com. Accessed 16 May 2013
12. S.J. McCarthy, G.F. Meijs, N. Mitchell, P.A. Gunatillake, G. Heath, A. Brandwood, K. Schindhelm, Biomaterials **18**, 1387 (1997)
13. D.J. Upadhyay, N.Y. Cui, C.A. Anderson, N.A. Brown, Colloids Surf. A Physicochem. Eng. Asp. **248**, 47 (2004)
14. T.G. Grasel, J.A. Pierce, S.L. Cooper, J. Biomed. Mater. Res. **21**, 815 (1987)
15. Y. Arima, H. Lwata, Biomaterials **28**, 2007 (2007)
16. T. M. Chapman, J. Polym. Sci. Part A: Polym. Chem. **27**, 1993 (1989)
17. A. Takahara, J. Tashita, T. Kajiyama, M. Takayanagi, Polymer **26**, 987 (1985)
18. K.G. Tingey, J.D. Andrade, Langmuir **7**, 2471 (1991)
19. T. Fujiwara, K.J. Wynne, Macromolecules **37**, 8491 (2004)

Temperature Dependence of Electrical Conductivity of Carbon Nanotube Films from 300 to 1100 K

Xiaoshan Zhang and Haitao Liu

Abstract Carbon nanotube (CNT) films are the most promising high-temperature electronic materials of the future. Studying the conductivity-temperature characteristic of the CNT films is an effective method to understand their nature of conduction. In this work, we present first measurement of the temperature dependence of electrical conductivity of the direct spun CNT film in a wide temperature range, from 300 up to 1100 K. The results indicate that the conductivity of the film increases with increase of the temperature up to the crossover temperature (T^*), above which it starts decreasing. We find that the conduction of the CNT film is determined by two components, the individual nanotube resistance (R_{tube}) and the contact resistance (R_{contact}). At the temperature below T^*, R_{contact} plays a major role in the charge transport. Above T^*, both R_{tube} and R_{contact} determine the conduction behavior, due to the electron–phonon scattering in the individual CNT at high temperatures. We propose that the study of conductivity-temperature characteristic of the CNT films provides lots of information for understanding the conduction mechanisms of the films and thus aids to improve their conductivity and applications at high temperatures.

1 Introduction

Carbon nanotube (CNT) films, i.e., macroscopic assemblies composed of aligned or randomly oriented CNT, have great potential for conducting and sensing applications owing to their low weight, excellent mechanical performance, and high electrical conductivity [1–6]. Theoretically, the CNT films composed of defectless individual single-wall nanotubes (SWNT), with precise length of contacts between them, can conduct as an armchair nanotube i.e., transport electrons ballistically (without scattering) [7]. Such films are superior to any traditional metallic conductors (e.g.,

X. Zhang · H. Liu (✉)
Science and Technology on Advanced Ceramic Fibers and Composites Laboratory,
College of Aerospace Science and Engineering, National University of Defense Technology,
Changsha, China
e-mail: htslht@163.com

Y. Han (ed.), *Physics and Techniques of Ceramic and Polymeric Materials*,
Springer Proceedings in Physics 216,
https://doi.org/10.1007/978-981-13-5947-7_20

aluminum and copper). They would be a promising conductive material for the next generation electrical wires. Moreover, the studies have shown that the CNT was stable in microstructure up to approximately 700 K in air or 2000 K in oxygen-free atmosphere [8, 9]. It indicates that the CNT films can replace the traditional conductors in high-temperature services (e.g., aerospace, nuclear, and engine etc.), where the traditional conductors get damaged or completely failed. Therefore, the understanding of the conduction mechanisms and electrical properties of the CNT films is one of the key topics for the use of CNT films in high-temperature fields. However, due to the difficulty in measuring the high-temperature conductivity of the CNT film. There are few studies have been made to investigate the temperature dependence of conductivity of the CNT film at high temperatures.

This work aims at casting some light into this issue by studying the temperature dependence of conductivity of the direct spun CNT films in a wide temperature range (300–1100 K). We find the conduction behavior was related to the temperature and the conduction mechanisms were presented to explain the temperature-conductivity characteristic of the CNT film. Meanwhile, this study can help for the development of conductivity of the CNT films and promotes the films that are widely used in high temperature fields.

2 Experimental

The CNT films (SCNC, produced in Suzhou Creative Nano Carbon Co., Ltd., China) were fabricated by the chemical vapor deposition method [1]. The microstructures of the CNT film were characterized by scanning electron microscopy (SEM, Helios 600i FEI). Transmission electron microscopy (TEM, Tecnai F20) was utilized to characterize the morphologies and microstructures of the CNT.

Before the conductivity measurements, the film samples were cut into ribbons with ~20 mm in length and ~2 mm in width. Then, the film sample was pasted on a ceramic plate by the high temperature conductive adhesive. They were placed in a continuous flow N_2 quartz tube furnace (OTL 1200); the temperature was slowly varied from 300 to 1100 K at a rate of ~10 K/min. Two Pt wires were used to connect two ends of the tested samples with the Keithley 2000 source meter. The electrical conductivity (σ) was calculated using the following formula:

$$\sigma = \frac{L}{Rwh} \tag{1}$$

where L, w, h, and R are the sample length, width, thickness, and resistance.

3 Results and Discussion

3.1 Microstructures of the CNT Film

Figure 1a, b show the surface and cross-sectional morphologies and microstructures of the CNT film. As shown in Fig. 1a, the as-received CNT film was composed of randomly oriented CNT bundles. A large number of micro-scaled pores were observed on the surface of the film. They were detrimental to the electrical conductivity of the CNT film [10]. As shown in Fig. 1b, the thickness of the CNT film was ~9 μm. Figure 1c shows the TEM microstructure of the CNT film. As shown in Fig. 1c, the as-received film was composed of multi-wall CNT with the diameter ~9 nm.

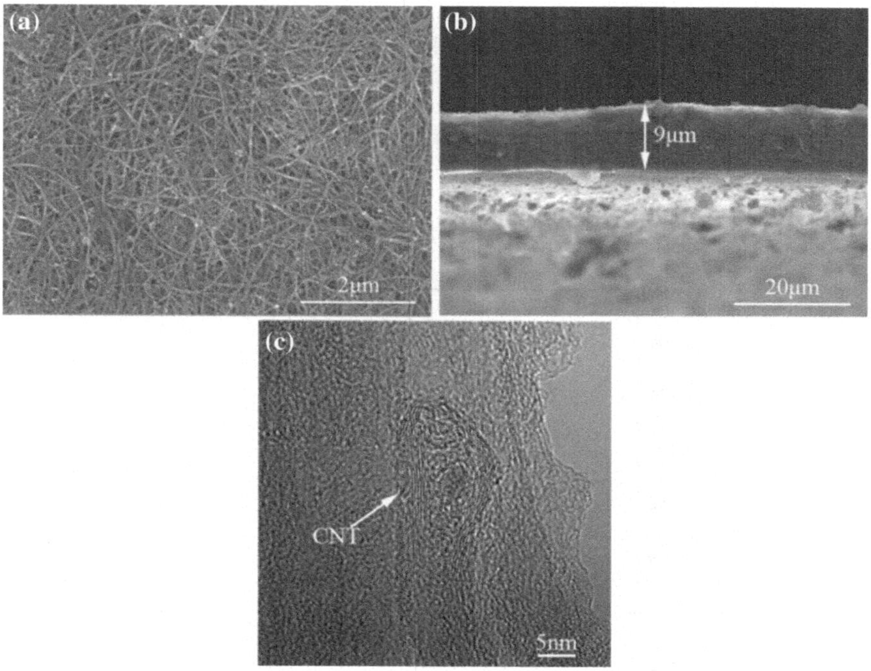

Fig. 1 **a** Surface morphologies and microstructures of the CNT film; **b** the cross-sectional microstructures of the CNT film; and **c** the TEM microstructure of the CNT film

3.2 Conductivity-Temperature Characteristic of the CNT Film

Figure 2 shows the temperature dependence of the electrical conductivity, $\sigma(T)$, of the CNT film between 300 and 1100 K. We may obviously distinguish two temperature ranges. Starting from 300 K, the conductivity increases with increasing temperature (range I, semiconductive behavior), and shows a broad maximum around 700 K. At the temperature above 700 K, the conductivity decreases monotonically with increasing temperature up to 1100 K (range II, metallic behavior).

The unusual temperature-conductivity characteristic indicates that the conductivity and conduction mechanisms of the film are highly dependent on temperature. In a macroscopic CNT film, where the film extends over length scales far greater than an individual nanotube, there are many junctions between the nanotubes within a film. These junctions create the contact potential barrier and the electrons ballistic-type transport within a nanotube is interrupted [11]. Therefore, unlike that of an individual CNT, the conductivity of the CNT film is determined by two components: the resistance of the individual CNT (R_{tube}) and the contact resistance between CNT (R_{contact}). The total resistance of the CNT film can be written as (2) [12]:

$$R = R_{\text{tube}} + R_{\text{contact}} \tag{2}$$

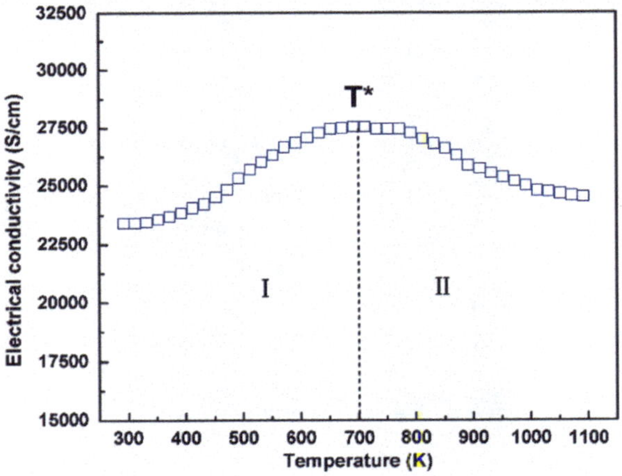

Fig. 2 Temperature dependence of electrical conductivity of the CNT film (the dashed line separates the two temperature ranges)

3.3 Temperature Dependence of the R_{tube}

It has been found that the R_{tube} is actually temperature dependent [13, 14]. Several experiments have demonstrated ballistically transport (i.e., without scattering) of the electrons in nanotubes at the temperature below the typical transition temperature (T_{c}, which depends on the diameter, walls, and chirality of the CNT) [13–16]. Therefore, the electrons freely propagate in the nanotubes and the R_{tube} is very small at the temperature below T_{c}. As the temperature increases, the electrons get excess energy due to heating the nanotubes significantly. When the electron has an excess energy higher than the phonon energy, the electrons rapidly emit the phonons. The created phonons can scatter (i.e., electron–phonon collisions) the electrons and set a limit on the maximal current transport of the CNT [12, 17]. The higher the temperature is, the stronger the scattering effect. It had been found that the temperature rises up to ~1000 K, even creating significant light emission [18]. This will cause a dramatic increase in R_{tube} above T_{c}. Other studies have also shown that the electron–phonon scattering considerably confined the conductivity of the CNT [19, 20].

3.4 Temperature Dependence of the $R_{contact}$

It has been found that the resistances of the CNT films were often two orders of magnitude higher than the individual CNT, which suggested that strong scattering occurs at the tube boundaries as a result of contact potential barrier (E_{c}) between the CNT [21, 22]. Studies have shown that the contact barrier height decreases with increasing temperature [23, 24]. Moreover, according to the theory of dielectric physics, the temperature dependence of contact conductivity between the CNT, $\sigma_{\mathrm{contact}}(T)$ decided mainly by hopping electrons between the CNT reads as [15]:

$$\sigma_{\mathrm{contact}}(T) = \frac{K_{\mathrm{h}}}{T}\exp\left(-\frac{U}{K_{\mathrm{b}}T}\right) \tag{3}$$

K_{h} is constant, U is the potential barrier, K_{b} is the Boltzmann constant, and T is absolute temperature. As expressed by (3), the conductivity caused by hopping electrons is enhanced by increasing temperature. The results indicate that the R_{contact} decreases with the increasing temperature.

3.5 Conduction Mechanism of the CNT Film

As the temperature increases, the R_{contact} gradually decreases, while the R_{tube} significantly increased when the temperature is above T_{c}. As a result, the conductivity of the films may show a maximum at a certain temperature depending on which a

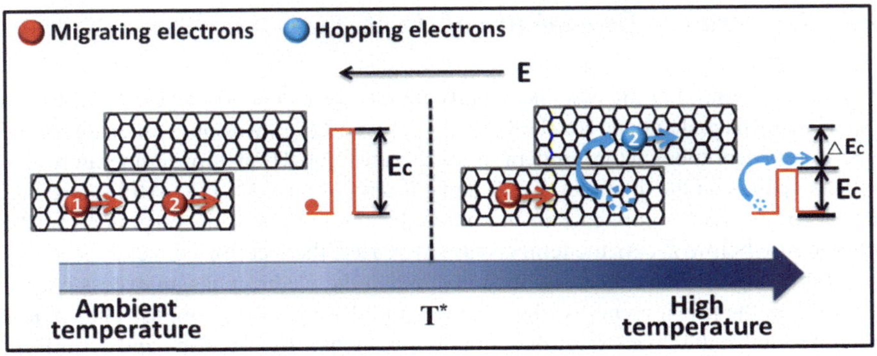

Fig. 3 Schematic illustrations of electron transportation in CNT and CNT bundles at different temperatures (ΔE_c: the decrease of contact potential barrier)

component plays a major role. We present the crossover temperature of the CNT film: $T^* \approx 700$ K (Fig. 2).

Figure 3 shows the electrons transport in the CNT and CNT bundle at different temperatures. At the temperature below T^*, the migrating electrons can freely propagate in the CNT. In this case, the R_{tube} is very small, the electronic conduction is mainly determined by the R_{contact}. With the temperature increasing, the hopping conductivity between the CNT increased due to the E_c height decreased. Therefore, the conductivity of the films increases with increasing temperature.

Above T^*, the thermal fluctuations contribute to enhancing the hopping conductivity between the CNT, while the electron–phonon scattering by the phonons limits the maximal current transport of the individual CNT. As a result, both the R_{contact} and R_{tube} determine the conduction of the electrons. It can be seen from Fig. 2, above T^*, the conductivity shows metallic behavior ($d_\sigma/d_T < 0$). It indicates that the R_{tube} play a major role in electronic conduction. The metallic behavior was caused by the electron–phonon scattering at high temperature, for which the maximal current transport of the CNT was significantly limited.

4 Conclusion

In conclusion, the temperature dependence of electrical conductivity of the direct spun CNT films was investigated in detail from 300 up to 1100 K. We have found that the conductivity and conduction mechanism of the films are determined by two components, R_{tube} and R_{contact}. The temperature dependence of the electrical conductivity follows the conduction mechanisms which take place in two different temperature ranges. Below the temperature T^*, the conductivity increases with the increasing temperatures. The conductivity enhanced mainly due to the decrease of R_{contact} with the increasing temperature. Above T^*, the conductivity shows an unexpected drop with the increasing temperature. The metallic behavior results from the electron–phonon scattering, which limits the conductivity of the CNT and leads to the significant increase of R_{tube}. All the results confirm that analyzing the conductivity-temperature characteristics may be highly useful to understand the conduction mechanisms of CNT film and their electrical properties at high temperatures.

References

1. M. Zhang, S.L. Fang, A.A. Zakhidov, S.B. Lee, A.E. Aliev, C.D. Williams, K.R. Atkinson, R.H. Baughman, Strong, transparent, multifunctional, carbon nanotube sheets. Science **309**, 1215–1219 (2005)
2. Z.C. Zhang, Y.Z. Gu, S.K. Wang, Q.W. Li, M. Li, Z.G. Zhang, Enhanced dielectric and mechanical properties in chlorine-doped continuous CNT sheet reinforced sandwich polyvinylidene fluoride film. Carbon **107**, 405–414 (2016)
3. I.W.P. Chen, R. Liang, H.B. Zhao, B. Wang, C. Zhang, Highly conductive carbon nanotube buckypapers with improved doping stability via conjugational cross-linking. Nanotechnology **22**, 485708 (2011)
4. J.N. Wang, X.G. Luo, T. Wu, Y. Chen, High-strength carbon nanotube fiber-like ribbon with high ductility and high electrical conductivity. Nat. Commun. **5**, 3848 (2014)
5. L. Dumee, K. Sears, S. Mudie, N. Kirby, C. Skourtis, J. Mcdonnell, S. Lucas, J. Schutz, N. Finn, C. Huynh, S. Hawkins, L. Kong, P. Hodgson, M. Duke, S. Gray, Characterization of carbon nanotube webs and yarns with small angle X-ray scattering: revealing the yarn twist and inter-nanotube interactions and alignment. Carbon **63**, 562–566 (2013)
6. J.T. Di, D.M. Hu, H.Y. Chen, Z.Z. Yong, M.H. Chen, Z.H. Feng, Y.T. Zhu, Q.W. Li, Ultrastrong, Foldable, and highly conductive carbon nanotube film. ACS Nano **6**, 5457–5464 (2012)
7. K. Liu, Y. Sun, X. Lin, R. Zhou, J. Wang, S. Fan, K. Jiang, Scratch-resistant, highly conductive, and high-strength carbon nanotube based composite yarns. ACS Nano **4**, 5827–5834 (2010)
8. Y.A. Kim, H. Muramatsu, T. Hayashi, M. Endo, M. Terrones, M.S. Dresselhaus, Thermal stability and structural changes of double-walled carbon nanotubes by heat treatment. Chem. Phys. Lett. **398**, 87–92 (2004)
9. Z.B. Yang, X.M. Sun, X.L. Chen, Z.Z. Yong, G. Xu, R.X. He, Z.H. An, Q.W. Li, H.S. Peng, Dependence of structures and properties of carbon nanotube fibers on heating treatment. J. Mater. Chem. **21**, 13772 (2011)
10. P. Liu, Y.F. Tan, D. Hu, D. Jewell, M.D. Hai, Multiproperty enhancement of aligned carbon nanotube thin films from floating catalyst method. Mater. Des. **10**, 754–760 (2016)
11. S. Ravi, A.B. Kaiser, C.W. Bumby, Charge transport in surfactant-free single walled carbon nanotube networks. Phys. Status Solidi B **250**, 1463–1467 (2013)
12. A. Jorio, G. Dresselhaus, M.S. Dresselhaus, Carbon nanotube. Topics Appl. Phys. **111**, 455–493 (2008)
13. G.M. Zhao, Evidence for room-temperature superconductivity in carbon nanotubes. New Res. Supercond. **1**, 23–59 (2002)
14. W. Liang, M. Bockrath, D. Bozovic, J.H. Hafner, H. Park, Fabry-Perot interference in a nanotube electron waveguide. Nature **411**, 665 (2001)
15. J. Kong, E. Yenilmez, T.W. Tombler, W. Kim, H.R.B. Laughlin, L. Liu, Quantum interference and ballistic transmission in nanotube electron waveguides. Phys. Rev. Lett. **87**, 106801 (2001)
16. A. Javey, J. Guo, Q. Wang, M. Lundstrom, H.J. Dai, Ballistic carbon nanotube field-effect transistors. Nature **424**, 654–657 (2003)
17. Z. Yao, C.L. Kane, C. Dekker, High-field electrical transport in single-wall carbon nanotubes. Phys. Rev. Lett. **84**, 2941–2944 (2000)

18. D. Mann, Y.K. Kato, A. Kinkhabwala, E. Pop, J. Cao, X.R. Wang, Electrically driven thermal light emission from individual single-walled carbon nanotubes. Nature Nanotechnol. **2**, 33–38 (2004)
19. J.Y. Park, S. Rosenblatt, Y. Yaish, V. Sazonova, H. Ustunel, S. Braig, Electron-phonon scattering in metallic single-walled carbon nanotubes. Nano Lett. **4**, 517–520 (2004)
20. A. Javey, J. Guo, M. Paulsson, Q. Wang, D. Mann, M. Lundstrom, High-field quasi ballistic transport in short carbon nanotubes. Phys. Rev. Lett. **92**, 106804 (2004)
21. W.A. Heer, W.S. Bacsa, A. ChAtelain, T. Gerfin, R. Humphrey-Baker, L. Forro, Aligned carbon nanotube films: production and optical and electronic properties. Science **268**, 845–846 (1995)
22. M. Zhang, K.R. Atkinson, R.H. Baughman, Multifunctional carbon nanotube yarns by downsizing an ancient technology. Science **306**, 1358–1361 (2004)
23. Y.J. Kim, T.S. Shin, H.D. Choi, J.H. Kwon, Y.C. Chung, H.G. Yoon, Electrical conductivity of chemically modified multiwalled carbon nanotube/epoxy composites. Carbon **43**, 23–30 (2005)
24. M. Salvato, M. Cirillo, M. Lucci, S. Orlanducci, I. Ottaviani, M.L. Terranova, Charge transport and tunneling in single-walled carbon nanotube bundles. Phys. Rev. Lett. **101**, 246804 (2008)

Study on the Semiconducting Grain and Insulating Barrier Layer in Aluminum/Niobium Co-doped CCTO

Ali Wen, Yanyan Zhang, Jiliang Zhu and Daqing Yuan

Abstract In this paper, aluminum/niobium co-doped calcium copper titanate ceramics were synthesized by a dry route based on the chemical formula of $CaCu_3Ti_{4-x}Al_{0.5x}Nb_{0.5x}O_{12}$ with $x = 0.0, 0.2, 0.5$, and 5.0%. The dielectric constants of the ceramics were over 10^4. X-ray diffraction and scanning electron microscopy results show the high dielectric constant comes from the capacitance effect of inner barrier layers (IBLC). In order to explore the origin of the semiconducting grains in it, X-ray photoelectron spectroscopy (XPS) was used. The experimental results show the existence of Ti^{3+} ions in sample, which caused the lattice polaronic distortion and the formation of Ti^{3+}–O–Ti^{4+} bonds. Under the applied electric field, the polaron can be transported from a Ti^{3+}–O–Ti^{4+} bonds to another, which leads to the generation of dc conduction. The existence of Ti^{3+} ion results in the semiconducting of grains in aluminum/niobium co-doped calcium copper titanate ceramics. The formation process of Ti^{3+} ion was also discussed.

1 Introduction

Dielectric materials are widely used to make capacitors, resonators, filters, etc. [1–4]. Dielectric materials with high dielectric constant can significantly decrease the size of capacitors. And for two capacitors with the same size, the energy storage

A. Wen · D. Yuan (✉)
China Institute of Atomic Energy, Beijing 102413, China
e-mail: yuandq@ciae.ac.cn

A. Wen
e-mail: aliwen89@163.com

Y. Zhang
Qingdao Technological University Qindao College, Qingdao 266106, China
e-mail: yanyanzhang_zz@163.com

J. Zhu
College of Materials Science and Engineering, Sichuan University, Chengdu 61064, China
e-mail: jlzhu167@scu.edu.cn

Y. Han (ed.), *Physics and Techniques of Ceramic and Polymeric Materials*,
Springer Proceedings in Physics 216,
https://doi.org/10.1007/978-981-13-5947-7_21

ability of the capacitor with high dielectric constant is larger. So for the demands of high-density energy storage and devices miniaturization, colossal dielectric materials have attracted researchers' extensive attention in recent years [5, 6]. At present, the main colossal dielectric materials are the $BaTiO_3$ and the relaxor ferroelectrics (e.g., $Pb(Mg_{1/3}Nb_{2/3})O_3$) [7, 8]. However, the dielectric constant of $BaTiO_3$ is quite unstable. And relaxor ferroelectrics, which contain lead, are not environmentally friendly. Therefore, calcium copper titanates (CCTO) with high dielectric constant (10^4–10^5) and temperature stability in 100–400 K are practically useful materials [8, 9].

However, the relatively high dielectric loss of calcium copper titanate ceramics limits its application. Our previous work studied the high-performance material of aluminum and niobium (AN) co-doped CCTO ceramics [10]. Compared with the undoped CCTO, at room temperature the dielectric loss of CCTO with AN co-doped on its Ti^{4+} site ($CCT(AN)_xO$) decreases significantly when the testing frequency is higher than 10^4 Hz, while their dielectric constant is still over 10^4. The Nb doping at the Ti^{4+} sites can increase the dielectric constant, and the addition of Al_2O_3 can result in the wide frequency range in high dielectric constant. Our work made use of both of them. Therefore, the high performance of $CCT(AN)_xO$ ceramics makes it possible to be applied in miniaturization of communication devices and high-density energy storage. It is commonly explained that the high dielectric constant comes from an internal barrier layer capacitor (IBLC). Namely, in polycrystalline CCTO ceramic, its conducting interior grains are surrounded by thin insulating grain boundaries (GBs) [11–14]. But the origin of its barrier layer and semiconducting grains is still controversial. Li et al. suggested that the barrier layers might have relationship with the numerous twin boundaries, while its semiconducting grains were related to the valence change of Cu ions [15]. Lei Zhang et al. showed that the grain's dc conduction was due to the existence of Ti^{3+} and the resulting polaron relaxation in CCTO. However, its insulating layers are assumed to be the grain boundaries [16]. So it is meaningful to study the barrier layers and semiconducting grains in AN co-doped CCTO. This paper will explore the origin of the barrier layers in AN co-doped CCTO by X-ray diffraction (XRD), scanning electron microscopy (SEM), and its semiconducting grains by X-ray photoelectron spectroscopy (XPS).

2 Experimental

AN co-doped CCTO ceramic samples were synthesized by a dry route, according to the chemical formula $CaCu_3Ti_{4-x}Al_{0.5x}Nb_{0.5x}O_{12}$ ($CCT(AN)_xO$) with $x = 0.0$, 0.2, 0.5, and 5.0%. $CaCO_3$, TiO_2, Cu, and high-purity Nb_2O_5, Al_2O_3 powders in a stoichiometric ratio were used as raw materials. Powders' synthesis was performed at 900 °C for 10 h, and pellets were sintered at 1100 °C for 10 h followed by a slow cooling process in the furnace. The diameter of the pellets was 10 mm, and its thickness was about 1 mm. For the detailed preparation process and measurement conditions, refer to the reference [10]. Besides, the element valence of sample

$CCT(AN)_xO(x = 0.2)$ was analyzed by XPS, and the XPS analysis was performed using a Thermo Scientific ESCALAB 250Xi in narrow scan survey mode, and monochromatized AlKa ($hv = 1486.6$ eV) X-ray source was used.

3 Results and Discussion

Figure 1 shows the dielectric spectra of samples. Their dielectric permittivities and losses were tested at room temperature, and the frequency range of the test was from 10^2 to 10^6 Hz. These results agree with the reference values [8, 17, 18] It can be seen that the dielectric performance of $CCT(AN)_xO$ ($x = 0.2\%$) is better. Although the dielectric permittivity of AN co-doped CCTO is smaller than that of un-doped CCTO, it is still more than 10^4. Besides, when the frequency is four orders of magnitude higher, their losses are smaller than that of un-doped one. Moreover, at high frequency of over 10^5 Hz, the loss of $CCT(AN)_xO$ ($x = 0.2\%$)(<0.48) decreases by about ten percent of the un-doped one (>1).

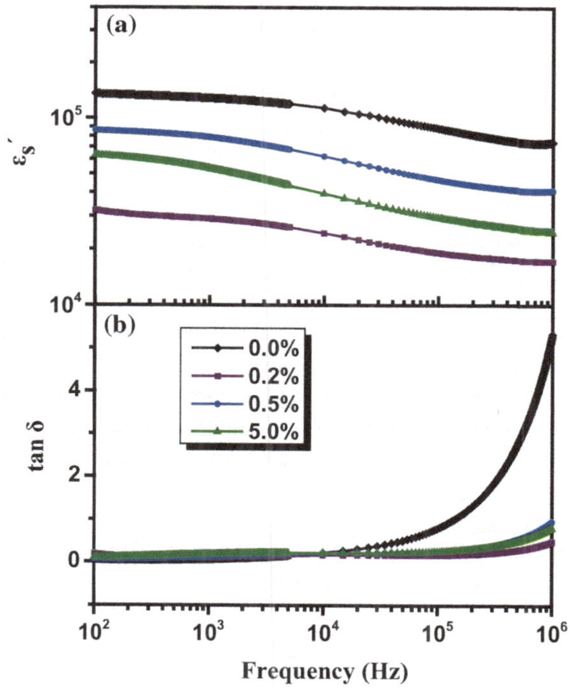

Fig. 1 Dependence of **a** the dielectric constants and **b** dielectric loss tangents with frequency at room temperature

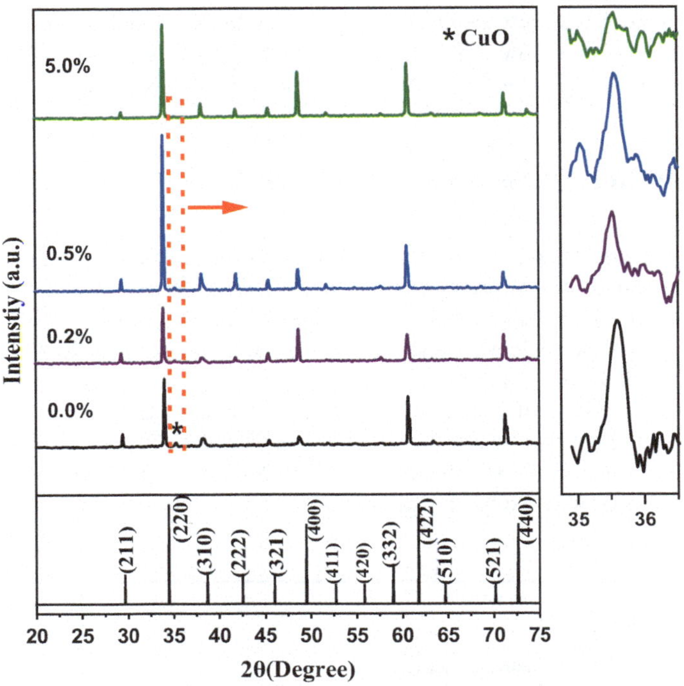

Fig. 2 XRD patterns of $CCT(AN)_xO$ ($x = 0.0, 0.2, 0.5$, and 5.0%) specimens

Figures 2 and 3 show the results of the XRD and SEM measurements of these samples, respectively. There are some weak CuO diffraction peaks in Fig. 2, and from Fig. 3, we can see that fields between grains are filled by another material, which is jam-like. The EDX measurement of the material shows that it mainly contains Cu and O with a ratio of 1:1 or 1:2. Figure 4 just shows the EDX point scan spectra of $CaCu_3Ti_4O_{12}$ ceramic grain and the grain boundary at these selected points. Therefore, the grains are surrounded by Cu-rich boundaries, which are consistent with the XRD results. So the high dielectric constant of AN co-doped CCTO ceramics comes from the extensively accepted IBLC effect, and these barrier layers are the composite of Cu-rich grain boundaries between the grains.

To assume that the thin GBs and the grains can form a capacitor, which has double dielectric layers, the capacitor's thickness is

$$\mathcal{D} = (\mathcal{D}_{gb} + \mathcal{D}_g). \tag{1}$$

One layer represents grain boundary, while the other represents grain. So here, $\mathcal{D}_{gb}$ is the thickness of GB, and $\mathcal{D}_g$ is grain size. Based on the assumption, samples' dielectric constant ε_s' can be obtained by

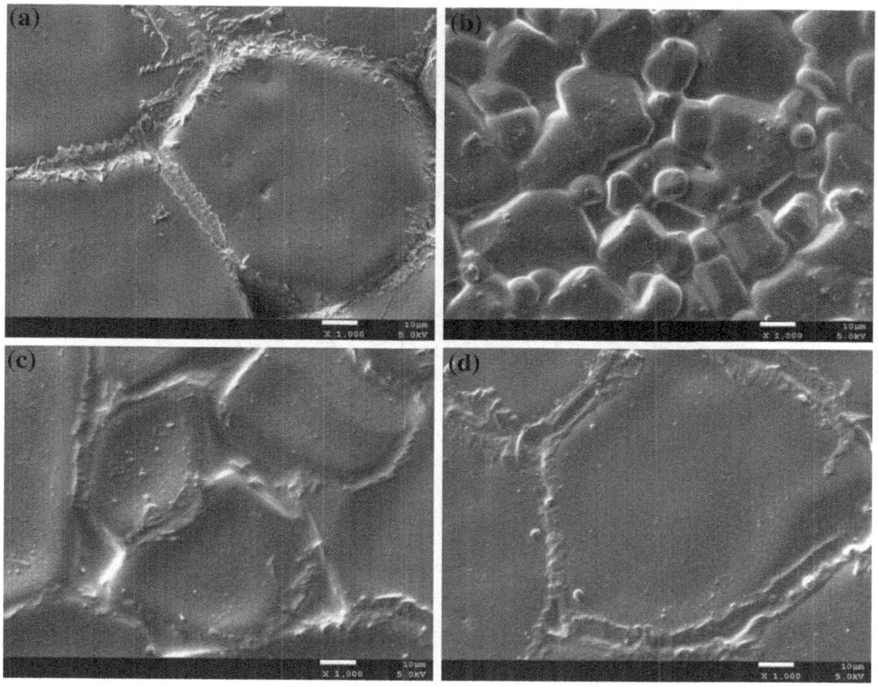

Fig. 3 SEM images from the unpolished surface of CCT(AN)$_x$O with **a** $x = 0.0\%$, **b** $x = 0.2\%$, **c** $x = 0.5\%$, and **d** $x = 5.0\%$

$$\varepsilon'_s = \varepsilon_{gb}(\mathcal{D}_{gb} + \mathcal{D}_g)/\mathcal{D}_{gb}, \tag{2}$$

where ε_{gb} is the GB dielectric constant [19]. When the ratio value of $\mathcal{D}/\mathcal{D}_{gb}$ is very large, even a small ε_{gb} may also result in a large ε'_s. As Fig. 3 shows, the thickness of GB is far smaller than the grain size. So the dielectric permittivities of AN co-doped CCTO ceramics are high. Grain boundaries play a leading role in AN co-doped CCTO ceramic's high dielectric permittivities.

As the reason for CCTO ceramic semiconducting grain formation, there are mainly three explanations as follows. One of them is the oxygen vacancy, which is due to the absence of oxygen in the sintering process [16]. The second is the valence state change of Ti^{4+} [16]. The third explanation is that the formation of the semiconducting grains is due to electron transitions between the Cu^{2+} and Cu^{3+} [20]. So it is necessary to test the valence of Ti, Cu, and O ions in AN co-doped CCTO and to study the reason of the semiconducting grains formation. To determine the ion valence variation, we did the XPS test of CCT(AN)$_x$O($x = 0.2\%$) ceramics, which show the best comprehensive dielectric properties among these CCT(AN)$_x$O ceramics.

Figure 5 shows the XPS peaks of the copper, titanium, aluminum, and oxygen and their fitting results. Figure 5a indicates that the binding energy difference of the

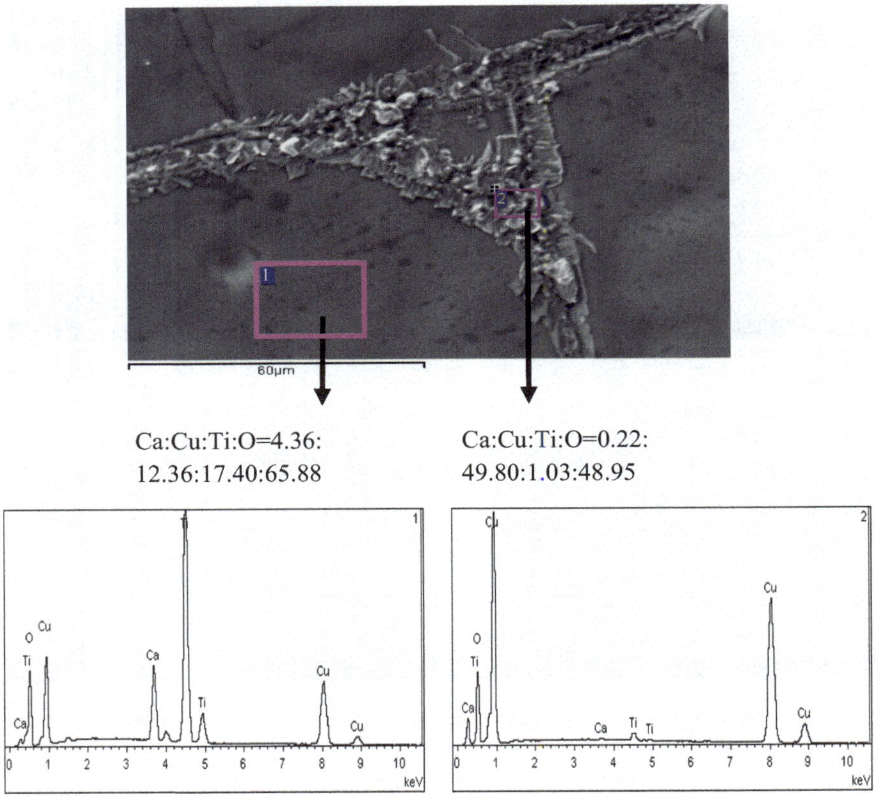

Fig. 4 EDX point scan spectra of (1) the grain and (2) GB of specimen $CCT(AN)_xO(x = 0.0\%)$ at selected fields

splitting peak of Cu 2p is 19.75 eV. The satellite peak position of Cu^{2+} is at 943 eV. The strong satellite peaks appearing at 942 and 961 eV indicate the 2^+ oxidation state of copper in the sample.

Figure 5b shows the binding energy difference of Ti 2p is 5.7 eV, and the peak at ~472 eV is the satellite peak of Ti 2p. The Ti $2p_{3/2}$ peak presents a dissymmetry. It can be fitted by two Gaussian–Lorentz curves. One curve represents the contributions of Ti^{3+}, and the other one is the contributions of Ti^{4+}, as shown by the mark in Fig. 5b. The positions of the two peaks obtained by fitting are 457.74 and 458.02 eV, respectively. According to the published literature, it is within the binding energy range of Ti^{3+} $2p_{3/2}$ and Ti^{4+} $2p_{3/2}$ [16].

Figure 5c shows the XPS spectrum of the Al 2p peak of $CCT(AN)_xO(x = 0.2\%)$ ceramic. Usually for aluminous XPS, their metal can have very obvious splitting peaks. The difference between the binding energy is 0.44 eV. However, it is difficult to observe the splitting peaks of its metal oxides. The Al 2p peak and Cu 3p peak always overlap as shown in Fig. 5c. The appearance of Cu 3p peak in Fig. 5c, which

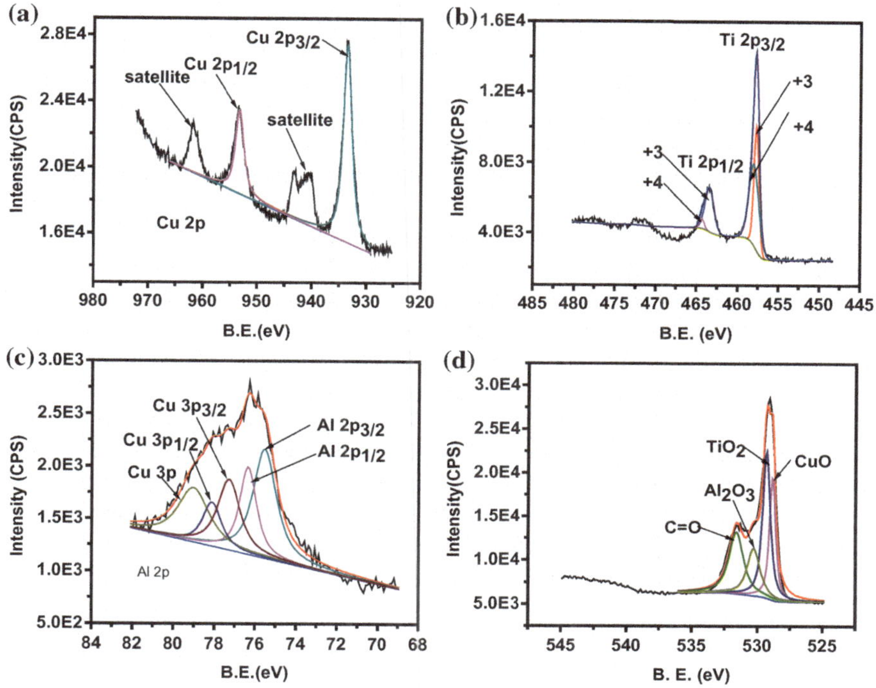

Fig. 5 XPS spectra of **a** Cu 2p, **b** Ti 2p, **c** Al 2p, and **d** O 1s of $CCT(AN)_xO$ ($x = 0.2\%$)

is attributed to CuO, also proves the oxidation state of copper in sample is the 2^+ state.

Figure 5d shows the XPS spectrum of the O 1s peak of $CCT(AN)_xO(x = 0.2\%)$ ceramic. As the binding energies of mental oxide are close, their XPS peaks may overlap when there exist various crystal forms. The fitted results of Fig. 5d show the composition of C=O. Considering the use of $CaCO_3$ raw material with decomposition temperature of 800 °C in sample preparation, they were decomposed completely during the longtime sintering of 10 h at the 1100 °C. The C=O composition in Fig. 5d may come from surface contamination.

In summary, the oxidation state of copper is the 2^+ state; Ti^{3+} and Ti^{4+} coexist in sample $CCT(AN)_xO(x = 0.2\%)$. Ti^{3+} ion has a radius of 0.670 Å, which is larger than that of Ti^{4+} (0.605 Å). So the existence of Ti^{3+} leads to polaronic distortion, and the Ti^{3+}–O–Ti^{4+} bonds could be formed. If the sample is under an applied electric field, the 3d electron in Ti^{3+} can jump to Ti^{4+}. As the Ti^{3+}–O–Ti^{4+} bonds form a linking path, the polaron can be transported through Ti^{3+}–O–Ti^{4+} bonds. Thus, dc conduction will occur. That is to say, these formations of semiconducting grains are mainly due to the Ti^{3+}/Ti^{4+} valence change.

The study by Li et al. suggests that some of Cu^{2+} in sample can be converted to Cu^{1+} at the high temperatures of sintering [15]. On cooling, some of the Cu^{1+} ions would be oxidized to Cu^{2+} and release an electron to the 3d band of Ti. This probably results in the existence of Ti^{3+} ions in sample CCT(AN)$_x$O($x = 0.2\%$). However, according to the study of Lei Zhang et al., the Ti^{3+} ions can be compensated by oxygen vacancies [16]. On cooling process, rapid re-oxidation may appear, when the cooling temperatures are above 1100 °C. But its actual stop temperatures are between 600 and 900 °C. As a result, the outer edge of each grain, to some extent, is well oxidized, while the grain's interiors are still oxygen deficient. So there are oxygen vacancies in grains interiors. As the effective charge of an oxygen vacancy is +2e, it can be neutralized by two 3d electrons of two Ti^{4+} and thus forms two Ti^{3+} ions for every oxygen vacancy. Oxygen vacancies are common in materials of perovskite [21–23]. So for CCTAN$_x$O, these existing Ti^{3+} ions could be compensated by oxygen vacancies. And the generation of oxygen vacancies is due to the lack of oxygen during the sintering process of grain at high temperature [13]. Because in sample CCT(AN)$_x$O($x = 0.2\%$), there is no peak of Cu^{1+} in the XPS spectrum of Cu 2p and the oxidation state of copper is 2^+ state; it is not the valence change of copper that results in the existence of Ti^{3+} in it. And the existing Ti^{3+} ions are compensated by these oxygen vacancies in grains. So the formation of semiconducting grains in sample is due to the lack of oxygen during the sintering process, and the Ti^{4+} ions valance change. This results in the existence of Ti^{3+} and the semiconducting grains.

4 Conclusions

AN co-doping can improve the dielectric performance of CCTO. According to the XRD and SEM results, the grains of the obtained sample are surrounded by Cu-rich boundaries. For AN co-doped CCTO ceramics, colossal dielectric constant comes from the IBLC model, which is generally accepted at present. XPS detects the existence of Ti^{3+} in sample of CCT(AN)$_x$O($x = 0.2\%$), which results in the formation of semiconducting grains. Thus, for AN co-doping CCTO ceramics, the barrier layer comes from the Cu-rich grain boundary, and the semiconducting grains show apparently dependent relationship with the valence change of Ti^{4+} ions and oxygen vacancies.

References

1. R.K. Pandey, W.A. Stapleton, J. Tate, A.K. Bandyopadhyay, I. Sutanto, S. Sprissler, S. Lin, Applications of CCTO supercapacitor in energy storage and electronics. AIP Adv. **3** (2013)
2. L. Ren, X. Zhao, L. Yang, K. Wu, Effect of CeO_2 and ZrO_2 doping on the dielectric characteristics of CCTO ceramics. In: 2017 IEEE Electrical Insulation Conference, EIC 2017, pp. 11–14 (2017)
3. R.T.A.R. Prasath, N.K. Roy, S.N. Mahato, P. Thomas, Mineral oil based high permittivity $CaCu_3Ti_4O_{12}$ (CCTO) nanofluids for power transformer application. IEEE Trans. Dielectr. Electr. Insul. **24**, 2344–2353 (2017)

4. Y. Rao, J. Yue, C.P. Wong, Material characterization of high dielectric constant polymer–ceramic composite for embedded capacitor to RF application. Mater. Sci. **92**, 2228–2231 (2004)
5. S. Marković, M. Lukić, Č. Jovalekić, S.D. Škapin, D. Suvorov, D. Uskoković, Sintering effects on microstructure and dielectric properties of CCTO ceramics (2012)
6. G. Riquet, S. Marinel, Y. Breard, C. Harnois, A. Pautrat, Direct and hybrid microwave solid state synthesis of $CaCu_3Ti_4O_{12}$ ceramic: microstructures and dielectric properties. Ceram. Int. 0–1 (2018)
7. A.J. Moulson, J.M. Herbert, Electroceramics: Materials, Properties, Applications, 2nd edn. (2003)
8. L. Singh, U.S. Rai, K.D. Mandal, N.B. Singh, Progress in the growth of $CaCu_3Ti_4O_{12}$ and related functional dielectric perovskites. Prog. Cryst. Growth Charact. Mater. **60**, 15–62 (2014)
9. A. Nautiyal, C. Autret, C. Honstettre, S. Didry, M.El Amrani, S. Roger, A. Ruyter, Dielectric properties of CCTO/$MgTiO_3$ composites: a new approach for capacitor application. Int. J. Adv. Nanomater. **1**, 27–40 (2015)
10. A. Wen, D.Q. Yuan, X.H. Zhu, J.G. Zhu, D.Q. Xiao, J.L. Zhu, Electrical and dielectric properties of aluminum/niobium co-doped $CaCu_3Ti_4O_{12}$ ceramics. Ferroelectrics **492**, 1–9 (2016)
11. R. Schmidt, M.C. Stennett, N.C. Hyatt, J. Pokorny, J. Prado-Gonjal, M. Li, D.C. Sinclair, Effects of sintering temperature on the internal barrier layer capacitor (IBLC) structure in $CaCu_3Ti_4O_{12}$ (CCTO) ceramics. J. Eur. Ceram. Soc. **32**, 3313–3323 (2012)
12. A. Nautiyal, C. Autret, C. Honstettre, S. De Almeida-Didry, M. El Amrani, S. Roger, B. Negulescu, A. Ruyter, Local analysis of the grain and grain boundary contributions to the bulk dielectric properties of $Ca(Cu_{3-y}Mg_y)Ti_4O_{12}$ ceramics: importance of the potential barrier at the grain boundary. J. Eur. Ceram. Soc. **36**, 1391–1398 (2016)
13. X.J. Luo, Y.S. Liu, C.P. Yang, S.S. Chen, S.L. Tang, K. Bärner, Oxygen vacancy related defect dipoles in $CaCu_3Ti_4O_{12}$: detected by electron paramagnetic resonance spectroscopy. J. Eur. Ceram. Soc. **35**, 2073–2081 (2015)
14. R. Schmidt, D. Sinclair, Capacitors. Theory of operation, behavior and safety regulations, in *$CaCu_3Ti_4O_{12}$ (CCTO) Ceramics for Capacitor Applications* (Nova Science Publishers, Inc., 2013)
15. J. Li, A.W. Sleight, M.A. Subramanian, Evidence for internal resistive barriers in a crystal of the giant dielectric constant material: $CaCu_3Ti_4O_{12}$. Solid State Commun. **135**, 260–262 (2005)
16. L. Zhang, Z.-J. Tang, Polaron relaxation and variable-range-hopping conductivity in the giant-dielectric-constant material $CaCu_3Ti_4O_{12}$. Phys. Rev. B **70** (2004)
17. S.-W. Choi, S.-H. Hong, Y.-M. Kim, Electric and dielectric properties of Nb-Doped $CaCu_3Ti_4O_{12}$ ceramics. J. Am. Ceram. Soc. **90**, 4009–4011 (2007)
18. S. Krohns, P. Lunkenheimer, S. Meissner, A. Reller, B. Gleich, A. Rathgeber, T. Gaugler, H.U. Buhl, D.C. Sinclair, A. Loidl, The route to resource-efficient novel materials. Nat. Mater. **10**, 899–901 (2011)
19. B. Shri Prakash, K.B.R. Varma, Microstructural and dielectric properties of donor doped (La^{3+}) $CaCu_3Ti_4O_{12}$ ceramics. J. Mater. Sci.: Mater. Electron. **17**, 899–907 (2006)
20. P. Thomas, K. Dwarakanath, K.B.R. Varma, Effect of calcium stoichiometry on the dielectric response of $CaCu_3Ti_4O_{12}$ ceramics. J. Eur. Ceram. Soc. **32**, 1681–1690 (2012)
21. R.A. Mackie, S. Singh, J. Laverock, S.B. Dugdale, D.J. Keeble, Vacancy defect positron lifetimes in strontium titanate. Phys. Rev. B—Condens. Matter Mater. Phys. **79**, 1–31 (2009)
22. H. Xiao, C. Yang, C. Huang, L. Xu, D. Shi, V. Marchenkov, I. Medvedeva, K. Baärner, Influence of oxygen vacancy on the electronic structure of $CaCu_3Ti_4O_{12}$ and its deep-level vacancy trap states by first-principle calculation. J. Appl. Phys. **111**, 063713 (2012)
23. S. Chikada, T. Kubota, A. Honda, S. Higai, Y. Motoyoshi, N. Wada, K. Shiratsuyu, Interactions between Mn dopant and oxygen vacancy for insulation performance of $BaTiO_3$. J. Appl. Phys. **120**, 1–6 (2016)

The Effect of Al Doping on Ferroelectric and Dielectric Properties of PLZT Transparent Electro-optical Ceramics

Bin Zhu, Zhaodong Cao, Xiyun He, Xia Zeng, Pingsun Qiu, Liang Ling and Suchuan Zhao

Abstract The PLZT (8.0/69/31) transparent ceramics with Al doping were prepared by hot-press sintering technique. A single perovskite structure was observed by X-ray diffraction in all samples. The ferroelectric and dielectric properties of the materials change obviously with increasing Al content: The maximum dielectric constant was raised to 16,000, while the dielectric constant (ε_r) at room temperature presented a decreasing tendency from 5764 to 2501. The polarization versus electric field (P-E) hysteresis loops change from antiferroelectric (AFE) phase to ferroelectric (FE) phase at room temperature. The grains of the PLZT ceramics grow up with less Al doping, but the grain size decreases, while the Al-doping content increases further. The PLZT ceramics with a little Al ($x = 0.50$ mol%) doping exhibit a better transparency and conspicuous variations in ferroelectric and dielectric properties. This will be helpful for designing the materials for optical modulators.

1 Introduction

The PLZT transparent ceramics were first reported in 1972 by Haertling [1]. Thence, more and more attentions have been received for PLZT ceramics because of their excellent dielectric, piezoelectric, and electro-optical properties.

B. Zhu · S. Zhao (✉)
The Science Academy, Shanghai University, 99 Shangda Road, Shanghai 200444, China
e-mail: zhaosc@staff.shu.edu.cn

B. Zhu
e-mail: zhubin@student.sic.ac.cn

B. Zhu · X. He (✉) · X. Zeng · P. Qiu · L. Ling
Shanghai Institute of Ceramics, Chinese Academy of Sciences, 1295 Dingxi Road, Shanghai 200050, China
e-mail: xyhe@mail.sic.ac.cn

Z. Cao
Shanghai Institute of Laser Plasma, China Academy of Engineering Physics, Chengzhong Road, Shanghai 201800, China

Y. Han (ed.), *Physics and Techniques of Ceramic and Polymeric Materials*, Springer Proceedings in Physics 216,
https://doi.org/10.1007/978-981-13-5947-7_22

Because of the obvious effects of composition on the ferroelectric, dielectric, and electro-optical properties, works have been carried out on the influences of compositions to the PLZT ceramics [2–4]. Some works were devoted to modify the performance of PLZT ceramics by ions doping [5–7]. Among them, PLZT (8.0/69/31) ceramics were found to present a unique electrically controlled light scattering performance. When the PLZT materials are applied to the appropriate electric field, an AFE phase to FE phase transition will be happened accompanied with the light scattering occurring. And the scattering FE phase can return automatically to the original transparent AFE phase when the applied electrical field is removed [4, 8]. These advantages make the PLZT (8.0/69/31) ceramics more feasible to optical device application. However, this electro-optical effect of the PLZT (8.0/69/31) ceramics, which is closely related to the material ferroelectricity, was found not large enough to the optical device application. So it is necessary to study the ferroelectric and dielectric properties of PLZT (8.0/69/31) ceramics modified by ions doping to optimize the electro-optical properties of the materials.

In this paper, a series of Al-modified PLZT ($Pb_{0.92}(La_{0.08-x}Al_x)(Zr_{0.69}Ti_{0.31})_{0.98}O_3$, $x = 0, 0.005, 0.010, 0.015$) ceramics were prepared by the conventional mixed oxide powder and hot-press sintering techniques. The effects of different Al components on the phase structure, microstructure, optical transmittance, dielectric and ferroelectric properties of PLZT transparent ceramics were experimented and analyzed systematically.

2 Experimental Procedure

PLZT (8/69/31) transparent ferroelectric ceramics with Al doping were prepared by the hot-press sintering techniques. 10 wt% of more PbO was added to compensate the lead evaporating in the high-temperature process of sintering. The main materials La_2O_3 (99.46%), PbO (99.7%), Al_2O_3 (91.75%), ZrO_2 (99.8%), and TiO_2 (99.99%) were ball milled for 4 h in the ethanol medium and then dried and sieved. Afterward, the mixed powders were pressed to cylindrical pellets. The pellets with different components of Al were sintered at 1250 °C for 16 h in oxygen atmosphere. Finally, the sintered samples were cut into appropriate size for measurements.

The phase structures of PLZT samples sintered were measured by Raman (DXR Raman Microscope, Thermo Nicolet, USA) and X-ray diffraction (XRD, UltimalV, Rigaku, Japan). The microstructures of the ceramics were obtained by scanning electron micrographs (SEM) (Hitachi S-3400N, Tokyo, Japan). The ferroelectric P-E hysteresis loops were measured using a workstation equipment (Radiant, Technologies, USA). The transmittance of the polished PLZT samples with Al doping was measured by an U2800 spectrophotometer (Hitachi, Tokyo, Japan). The dielectric properties of PLZT ceramics were measured as a function of temperature by using a HP4284A LCR meter (Hewlett-Packard, Palo Alto, CA).

3 Results and Discussion

Phase and Microstructure. The XRD patterns of PLZT ceramics (8.0/69/31) sintered by the hot-pressing techniques with different Al contents ($x = 0$, 0.50, 1.00, 1.50 mol%) are shown in Fig. 1. These XRD patterns indicate that all samples with different Al contents are pure perovskite structure without any secondary phase.

The Raman spectroscopy of the surface for PLZT samples was shown in Fig. 2. The measurements were taken at room temperature. It is analyzed by spectroscopy that Al doping changed the construction of ceramic grains [9].

The SEM images of fracture surfaces for PLZT (8.0/69/31) ceramics with Al doping are showed in Fig. 3, exhibiting dense microstructures with well-developed grains. With the Al-doping content (0, 0.50, 1.00, 1.50 mol%) increasing, the grain sizes varied obviously. It is showed that little Al doping promotes the grain growth of

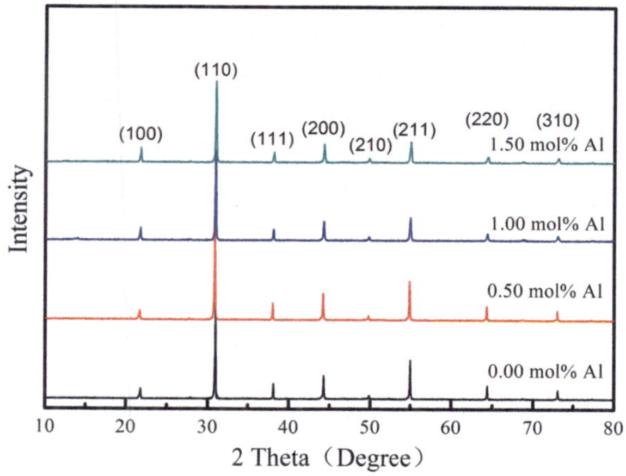

Fig. 1 Patterns of PLZT (8.0/69/31) ceramics with different components of Al doping

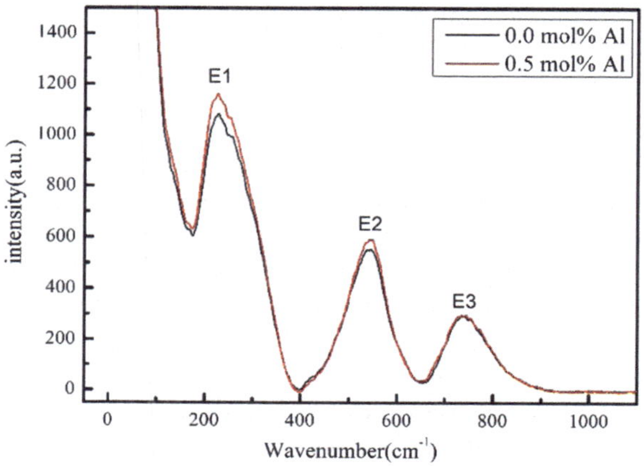

Fig. 2 Raman spectroscopy of PLZT (8.0/69/31) ceramics with different Al contents

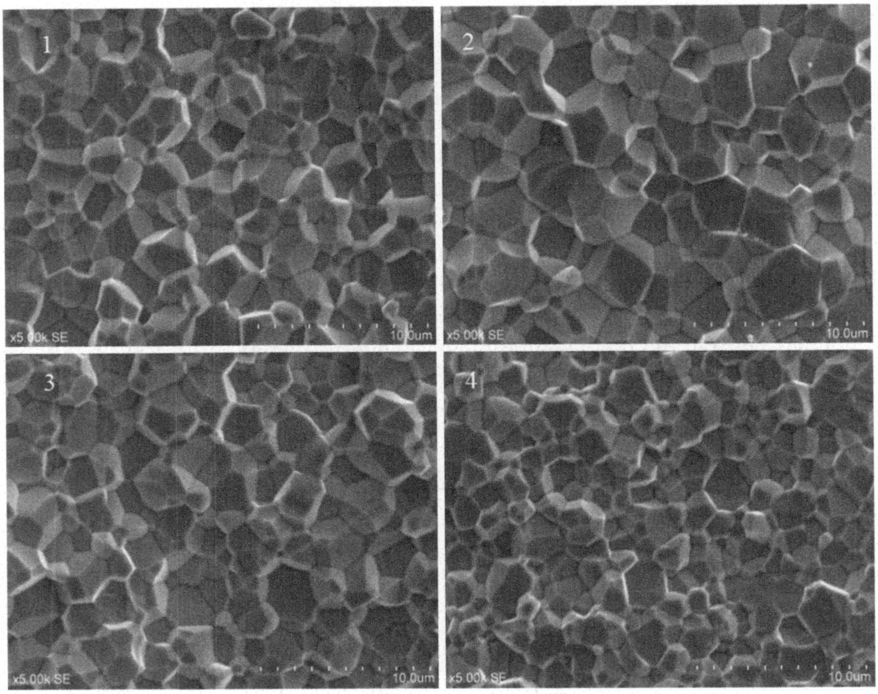

Fig. 3 SEM images of fracture surfaces for PLZT (8.0/69/31) ceramics with different components of Al doping: (1) $x = 0.00$ mol% Al, (2) $x = 0.50$ mol% Al, (3) $x = 1.00$ mol% Al, (4) $x = 1.50$ mol% Al

PLZT (8.0/69/31) ceramics. With Al content increasing continuously, the grain size of Al-modified PLZT samples decreases obviously. Some theoretical explains about grain growth were proposed [10]. The explanation is suggested as that, when the Al ion which has a smaller ionic radius (0.54 Å) doped into the crystal lattice, the lattice became loose, and this is benefit to the material diffusion transfer to promote the grain growth. But when more Al doped in PLZT materials, some Al would move into the grain boundary. The boundary with Al would resist the grain boundary migration and grain growth.

Optical Transmittance. Figure 4 shows the transmittance spectra of Al-modified PLZT (8.0/69/31) ceramic samples as a function of wavelength from 200 to 1100 nm with a thickness 0.50 mm. It can be found that a small amount of Al doping can improve the transmittance of the PLZT (8.0/69/31) ceramics in spectra. With Al-doping content increasing, the transmittance value of Al-modified PLZT samples dropped obviously.

Ferroelectric and Dielectric Properties. The P-E hysteresis loops of Al-modified PLZT (8.0/69/31) ceramics at room temperature are shown in Fig. 5. A typical antiferroelectric hysteresis loop is revealed for no Al doping, while other square

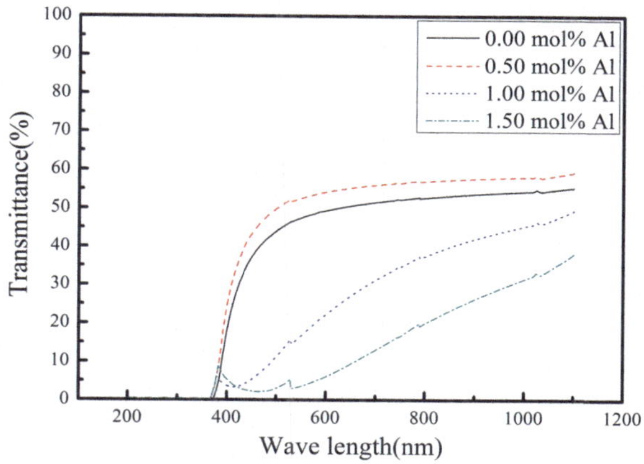

Fig. 4 Transmittance versus wavelength of Al-modified PLZT ceramics (thickness: 0.50 mm)

Fig. 5 P-E hysteresis loops of Al-modified PLZT ceramics at room temperature

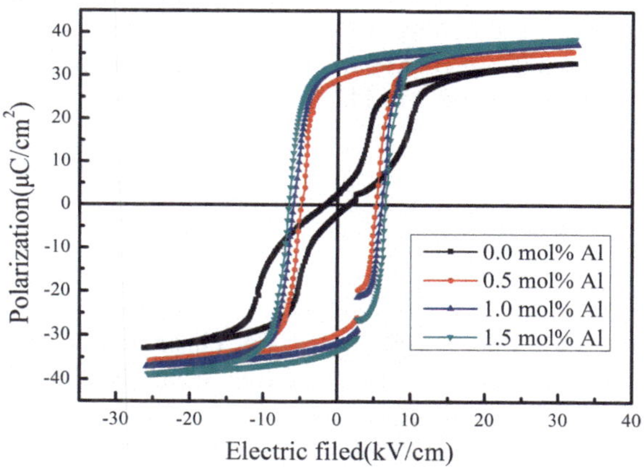

hysteresis loops are observed for Al-modified PLZT. The square hysteresis loops all show relatively large values of coercive electric field (E_c) and remnant polarization (P_r), which significantly indicate the normal ferroelectric characteristics [11]. And as the Al content increasing, the P_r value increased either while the E_c reached maximum.

In order to study the dielectric properties of PLZT ceramics, the dielectric constant ε_r is calculated by the following formula:

$$\varepsilon_r = Ct/(A\varepsilon_0) \quad (1)$$

where C is the capacitance value, t is the thickness of PLZT samples, A is the area of polished materials section, and ε_0 is the vacuum dielectric constant (8.8538×10^{-12} F/m). Changes of the dielectric constant for all Al-modified PLZT (8.0/69/31)

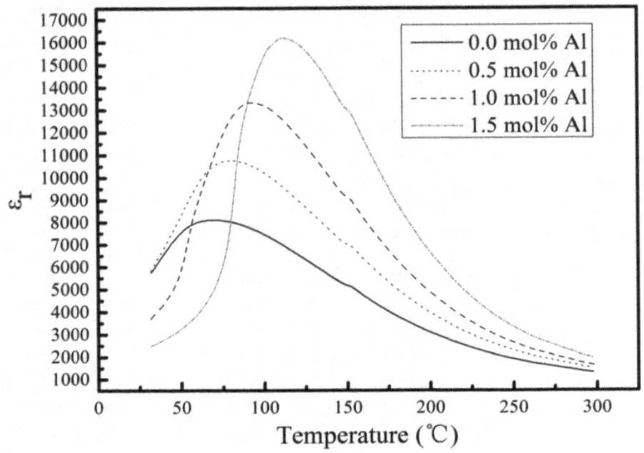

Fig. 6 Dielectric constant of Al-modified PLZT ceramics as a function of temperature at 1 kHz frequency

samples as a function of temperature at 1 kHz frequency are shown in Fig. 6. A broad diffused phase transition is discovered [12]. As the Al content increasing, the temperature corresponding to the respective maximum dielectric constant (T_m) increases from 68.9 to 112.0 °C.

4 Summary

Al-modified PLZT (8.0/69/31 + x mol% Al, $x = 0.00, 0.50, 1.00, 1.50$) transparent ceramics with a single perovskite phase and dense microstructure were prepared by conventional hot-press sintering techniques. With increasing Al-doping content, the maximum dielectric constant increases and T_m has a tendency moving from 69 to 112 °C, and the dielectric constant at room temperature (~32 °C) indicates a decreasing variation from 5764 to 2501. The P-E hysteresis loops change from antiferroelectric phase to normal ferroelectric phase with Al doping. A little Al doping can improve the transmittance of PLZT (8.0/69/31) ceramics. Doping a little Al in PLZT (8.0/69/31) ceramics can exert conspicuous influence on the ferroelectric and dielectric properties of the ceramics.

Acknowledgements This work was supported by the National Natural Science Foundation of China (Grant No. 51602327).

References

1. G.H. Haertling, C.E. Land, Hot-pressed (Pb, La)(Zr, Ti) O_3 ferroelectric ceramics for electrooptic applications. J. Am. Ceram. Soc. **54**(1), 1–11 (1971)

2. A. Kumada, G. Toda, Y. Otomo, An electrooptic image storage device using field induced reversible transitions between AFE and FE phases of PLZT ceramics. Ferroelectrics **7**(1), 367–369 (1974)
3. X. Zhang, B. Xia, X. Zeng et al., Temperature dependence of electric-induced light scattering performance for PLZT ceramics. J. Am. Ceram. Soc. **97**(5), 1389–1392 (2014)
4. B. Xia, X. He, X. Zheng et al., Electrically induced light scattering performances of lanthanum-modified lead zirconate titanate transparent ceramics. J. Electroceram. **29**(3), 192–197 (2012)
5. R. Rai, S. Sharma, R.N.P. Choudhary, Structural and dielectric properties of Bi modified PLZT ceramics. Solid State Commun. **133**(10), 635–639 (2005)
6. A. Limpichaipanit, A. Ngamjarurojana, Effect of Li and Bi co-doping and sintering temperature on dielectric properties of PLZT 9/65/35 ceramics. Ceram. Int. **43**(5), 4450–4455 (2017)
7. S. Somwan, A. Ngamjarurojana, A. Limpichaipanit, Dielectric, ferroelectric and induced strain behavior of PLZT 9/65/35 ceramics modified by Bi_2O_3 and CuO co-doping. Ceram. Int. **42**(9), 10690–10696 (2016)
8. B. Xia, X. He, D. Sun et al., The electrically controlled light scattering performances of PLZT transparent ceramics. Ceram. Int. **41**, S246–S249 (2015)
9. Y. Zhou, W. Zhao, W. Ruan et al., Abnormal electric-field-induced light scattering in Pb $(Mg_{1/3}Nb_{2/3})$ O_3-$PbTiO_3$ transparent ceramics. Appl. Phys. Lett. **106**(12), 122904 (2015)
10. R.A. Langman, R.B. Runk, S.R. Butler, Isothermal grain growth of pressure-sintered PLZT ceramics. J. Am. Ceram. Soc. **56**(9), 486–488 (1973)
11. L. Nie, X. He, C. Chang et al., Effect of anisotropy on the ferroelectric, optical, and electro-optic properties of PLZT transparent ceramics prepared by uniaxial hot-press sintering techniques. Opt. Mater. Express **6**(11), 3565–3573 (2016)
12. J. Zeng, G. Li, W. Ruan et al., The influence of domain structure on the optical and electrical properties of transparent (Pb, La)$(Mg_{1/3}Nb_{2/3})$ O_3–$PbTiO_3$ ceramics. Ceram. Int. **39**, S31–S34 (2013)

Preparation and Properties of PMMA Nanofibers with Photochromic and Photoluminescent Functions

Congcong Li, Peng Xi, Tianxiang Zhao, Xiaoqing Wang and Xuhuan Yan

Abstract A new polymer with spiropyran photochromic units (SP-PMMA) was synthesized by atom transfer radical polymerization (ATRP). The ^{1}H NMR, FT-IR, GPC and TGA results show that the SP-PMMA has good structure features such as controllable molecular weight, narrow molecular weight distribution, and good thermal stability. SP-PMMA nanofibers with multi-base photochromic and photoluminescent functions have been successfully prepared by electrospinning technique and synergistic coordination of the SP-PMMA with organic rare earth complexes. Scanning electron microscopy (SEM) observations show that the as-prepared nanofibers have a smooth surface, with the average diameter about 600 nm. Fluorescent and UV absorption spectra indicate that the SP-PMMA nanofibers can show red, green, yellow, blue-purple, and white colors under the excitation of 295, 367 nm UV lights and far-infrared light. The results of the present study verify that the SP-PMMA nanofibers can exhibit multi-base photochromic and photoluminescent functions. These kinds of multi-base photochromic and photoluminescent functions are of great significance for the development of multifunctional photochromic fibers.

1 Introduction

The photochromic phenomenon refers to that compound A is converted into B by isomerization reaction when irradiated with light of a certain wavelength, while compound B can be converted to A under the irradiation of another wavelength

C. Li · P. Xi (✉) · T. Zhao · X. Wang · X. Yan
Tianjin Polytechnic University, 399 Bin Shui West Road, Tianjin 300387, People's Republic of China
e-mail: xpsyq0007@sina.com

C. Li · P. Xi · T. Zhao
Tianjin Key Laboratory of Advanced Fibers and Energy Storage, Tianjin 300387, People's Republic of China

P. Xi
State Key Laboratory of Separation Membranes and Membrane Processes, Tianjin 300387, People's Republic of China

Y. Han (ed.), *Physics and Techniques of Ceramic and Polymeric Materials*,
Springer Proceedings in Physics 216,
https://doi.org/10.1007/978-981-13-5947-7_23

of light or heat [1, 2]. Under external stimuli, two states can be reversibly converted to each other under different wavelength lights. In general, with the progress of reversible photoisomerization reaction, the chemical structures of A and B will undergo a great change, and their corresponding photochemical and photophysical properties will also be changed, such as the absorption spectrum of the molecule [3], fluorescence spectrum [4], degree of conjugation, conductivity, magnetic properties, electrochemical properties, coordination ability, polarity, refractive index, permittivity, and hydrophilicity; these properties make the photochromic materials applied to the fields of information storage [5], optical switches [6], optical conversion devices [7], optical lenses [8], and photochromic fiber and fabric [9]. In recent years, with the development of society and technology, photochromic materials have also become the focus of academic attention [10]. Researchers have also made a series of achievements in the development of photochromic materials, which includes both organic and inorganic photochromic materials [11].

There are many kinds of organic photochromic materials, such as spiropyran [12], schiff base [13], and diaromatic ethylene [14]. Of these, spiropyran has been widely applied due to good photochromic function. The spiropyran molecules can realize the transformation from the closed-loop colorless spiropyran (SP) structure to the open-loop colored Merocyanine (MC) under the excitation of ultraviolet light [15]. As with photochromic materials, photoluminescent materials also have good photoluminescent functions. As an important member of the photoluminescent materials, the rare earth complexes have absorbed much attention because the rare earth ions have the function of high coordination [16] and the coordination number is usually 8–10. After coordination reaction of rare earth ions with organic ligands, the rare earth complexes with high fluorescence intensity can be obtained. At present, the application of these two kinds of photochromic and luminescent materials on fiber fabrics is mainly through simple doping. The obtained fabric is usually of single color, which cannot meet the demand of the colorful fabric market.

In this research, we prepared SP-PMMA nanofibers containing spiropyran photochromic units. Firstly, ATRP initiator containing spiropyran (SP-Br) was synthesized. And then, methyl methacrylate (MMA) was added, and the polymer with photochromic function (SP-PMMA) was prepared by ATRP method. With the SP-PMMA, the multi-base photochromic and photoluminescent nanofibers were prepared by the polymer molecular structure design, coordination reaction of the polymer with organic rare earth complex and electrospinning technique. The corresponding tested results show that the SP-PMMA has narrow molecular weight distribution and good spinning properties. The prepared SP-PMMA nanofibers have good multi-base photochromic and photoluminescent functions.

2 Materials and Methods

2.1 Materials

2-Bromoisobutyryl bromide (BIBB, Analytical grade) was purchased from Shanghai Aladdin Biochemical Technology Co., Ltd. Triethylamine (TEA, analytical grade), dichloromethane (DCM, analytical grade), copper (I) bromide (CuBr, analytical grade) and petroleum ether (PE, analytical grade), ethyl acetate (EAC, analytical grade) were obtained from Tianjin Sailboat Reagent Co., Ltd. Toluene, *N*,*N*,*N*,*N*,*N*-pentamethyldiethylenetriamine (PMDETA, analytically grade) was purchased from Tianjin Heowns Biotechnology Co., Ltd. Tetrahydrofuran (THF, analytical grade), *N*,*N*-dimethylformamide (DMF, analytical grade), and methyl methacrylate (MMA, analytical grade) were obtained from Tianjin Kemiou Chemical Reagent Co., Ltd. *N*-hydroxyethyl-3,3-Dimethyl-6-nitroindoline spiropyran (SP-OH) was from Sigma-Aldrich. The rare earth complex [$Tb(BA)_3$(phen), BA = benzoic acid, phen = 1,10-phenanthroline] was prepared as reported literature [17]. All reagents were dried and purified according to standard process before they were used.

2.2 Synthesis of SP-PMMA

ATRP initiator containing spiropyran (SP-Br) was prepared as reported by Claudia and Ventura [18–20]. SP-PMMA was prepared according to the following steps (Fig. 1). In a Schlenk bottle, the SP-Br (0.118 g, 0.235 mmol), PMDETA (98 μL, 0.469 mmol), MMA (5 mL, 47.093 mmol), Toluene (5 mL, 46.99 mmol) were added. The system was degassed by three freeze-pump-thaw cycles. And then, CuBr (0.034 g, 0.237 mmol) was added quickly at the last frozen state. The sealed flask was put into an oil bath at 60 °C under nitrogen gas. The reaction process was exposed to the air to stop. The as-prepared product was purified by precipitation through the addition of excess methanol and vacuum filtration step. The above step was repeated several times until the supernatant become colorless. The final product was dried at room temperature under vacuum condition.

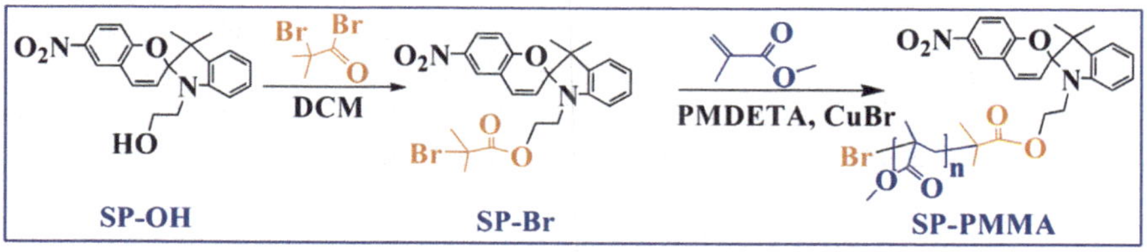

Fig. 1 Synthesis route of SP-PMMA

2.3 Preparation of Multi-base Photochromic SP-PMMA Nanofibers

The SP-PMMA nanofibers with photochromic and photoluminescent functions were prepared through electrospinning technology [21]. The appropriate amount of SP-PMMA, organic rare earth complex, and DMF was mixed to prepare the spinning solution. The prepared spinning solutions were placed in 10 mL plastic syringes, which had 22 G stainless-steel needles attached (inner diameter of 0.41 mm). Through a series of experiments, the optimal spinning process was found, the optimal spinning concentration was 28% ($mass_{SP\text{-}PMMA}$: $mass_{DMF}$ = 28%) the voltage was 20 kV, and the solution flow rate was 0.5 mL/h by using an injection pump, the distance between the tip of the needle and the collecting aluminum plate was 20 cm. The spinning temperature was 25 °C. The humidity was 30%. The samples, which were removed from collection devices and put in vacuum oven at 80 °C for the further characterization.

3 Results and Discussion

3.1 The Structure Characterization of SP-PMMA

As shown in the FT-IR spectra (Fig. 2), in the curve of SP-OH [22], the peaks at 1610 cm^{-1} and 1457 cm^{-1} are characteristic stretching vibration peaks of benzene rings. The infrared absorption peaks at 2960 cm^{-1} and 2865 cm^{-1} correspond to the $-CH_3$ and $-CH_2-$ stretching vibration, respectively. The –OH stretching vibration peak is located at 3280 cm^{-1}. In the curve of SP-Br, it can be seen from the peaks at 1733 cm^{-1} belonging to ester C=O characteristic peak, which showed that initiator SP-Br was prepared successfully. In the FT-IR spectrum of SP-PMMA, C=O characteristic peak of ester at 1733 cm^{-1} still exists. The $-CH_2-$ stretching vibration peaks at 2865–2960 cm^{-1} is enhanced. The results proved that SP-PMMA has been prepared successfully.

The structure of the SP-PMMA was further characterized through ^1H NMR [23] (Fig. 3). The chemical shift at 8.10 ppm, 7.94 ppm, 7.02 ppm, 6.72 ppm, 6.52 ppm, and 6.06 ppm (a–f) corresponds to the peaks of benzene ring. The chemical shift at 4.31 (g), 3.63 (h), and 1.91 ppm (i) belongs to the proton peaks of $-CH_2-$, respectively. The chemical shift at 3.68, 1.43 and 1.34 ppm is $-CH_3$ proton peaks at l, m, and j. The chemical shift at 1.29 ppm corresponds to the $-CH_3$ proton peak at k. Based on the above analysis, the SP-PMMA has been prepared successfully.

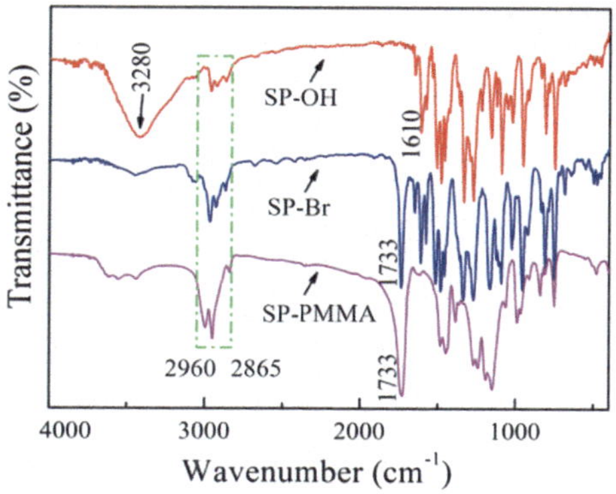

Fig. 2 FT-IR spectra of **a** SP-OH; **b** SP-Br; **c** SP-PMMA

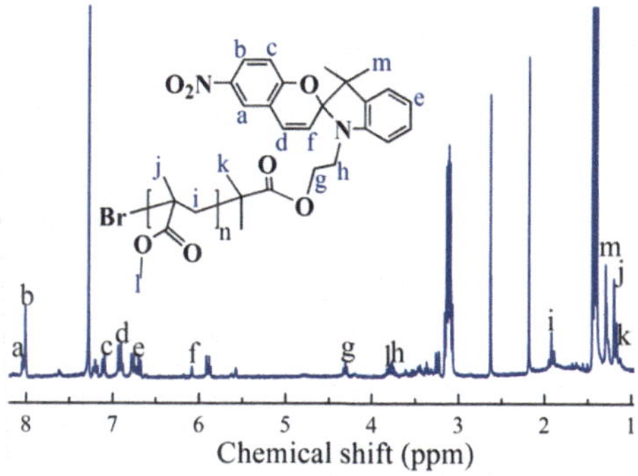

Fig. 3 ^{1}H NMR spectra of SP-PMMA

3.2 *Thermal Performance Analysis of SP-PMMA*

Figure 4 shows the TG curves of PMMA and SP-PMMA, from which we can see that the introduction of photochromic segment has little effect on the thermal stability of PMMA, and the decomposition temperatures of PMMA is basically same as that of SP-PMMA. Figure 5 shows the DSC curves of PMMA and SP-PMMA. It can be seen from Fig. 5 that the glass transition temperature (T_g) of SP-PMMA has a significant decrease compared with PMMA, which is due to the introduction of the photochromic segment increases the free volume between the polymer molecular chains, so that when the polymer molecules are heated, the segments can move freely at lower temperatures, and the T_g is lower, which is important for improving the toughness of fiber and nonwoven products.

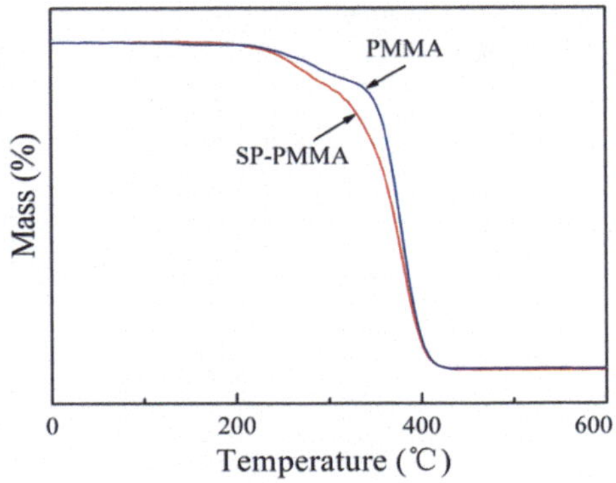

Fig. 4 TGA curves of PMMA and SP-PMMA

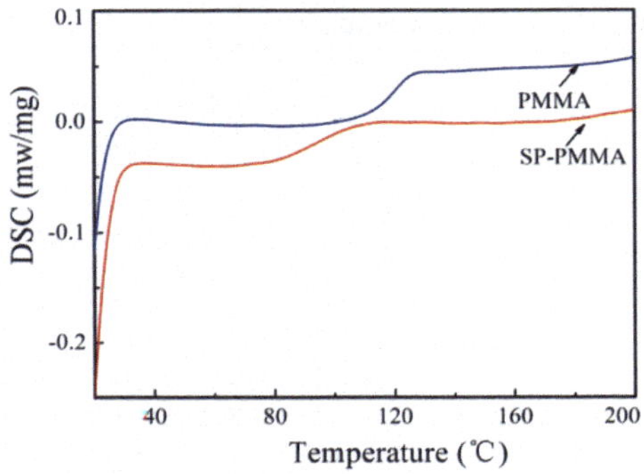

Fig. 5 DSC curves of PMMA and SP-PMMA

3.3 Gel Permeation Chromatography (GPC) Analysis of SP-PMMA

Figure 6 is the GPC analysis of SP-PMMA, the corresponding GPC trace is monomodal and narrow, which proves the main product is SP-PMMA. The product molecular weight is 48,054 g/mol, and the molecular weight distribution is 1.283 L, which provides the premise for spinning nanofiber.

3.4 Measurement and Characterization of SP-PMMA Nanofibers

3.4.1 Effects of the Applied Voltage on SP-PMMA Nanofiber Morphology

Figure 7 shows the SEM images of SP-PMMA nanofibers at different spinning voltages (a, b, c, and d images correspond to the spinning voltage 10, 15, 20, and

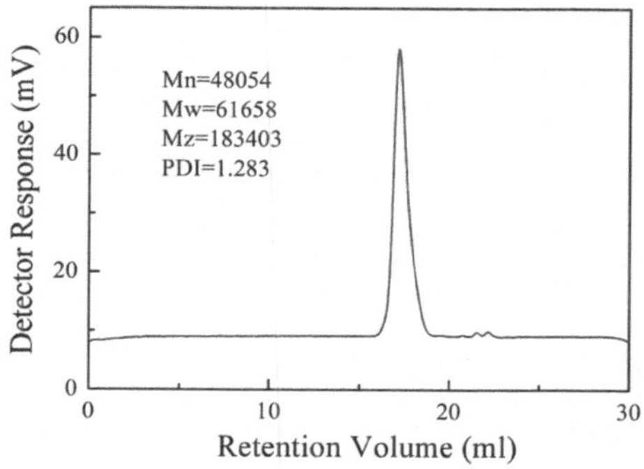

Fig. 6 GPC trace of SP-PMMA

Fig. 7 SEM images of SP-PMMA nanofibers at different voltages: **a** 10 kV; **b** 15 kV; **c** 20 kV; **d** 25 kV

25 kV, respectively). The surface of fibers is the smoothest and the diameter of the fiber is the most uniform when the voltage is 20 kV. So 20 kV is selected as the optimal spinning voltage [24].

3.4.2 Effects of the Receiving Distance Between the Tip and Collector on SP-PMMA Nanofiber Morphology

Figure 8 shows the SEM photographs at different receiving distances, which mean the distance between the tip of the needle and the collecting aluminum plate. The solvent in the spinning solution will not be able to volatilize when the receiving distance is 10 cm, and many small droplets are obtained on the receiving plate. It can be clearly seen from Fig. 8b that the fibers are more regular and have a uniform

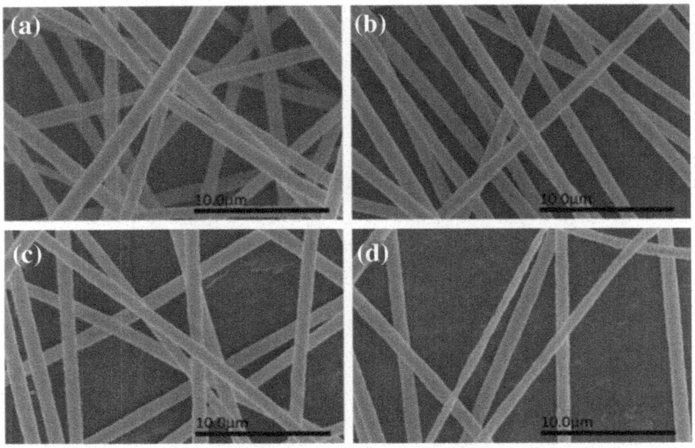

Fig. 8 SEM images of SP-PMMA nanofibers at different receiving distances **a** 15 cm; **b** 20 cm; **c** 25 cm; **d** 30 cm

size distribution when the receiving distance is 20 cm. And 20 cm is selected as the optimal receiving distance [25].

3.4.3 Effects of the Propulsion Speed on SP-PMMA Nanofiber Morphology

Figure 9 shows the images of SP-PMMA nanofibers under different propulsion speeds. By comparing the SEM images under different propulsion speeds, we can see that the fiber surface was smooth and the size distribution was uniform when spinning speed at 0.5 mL/h, which is the best spinning speed [26].

3.5 Morphologies of SP-PMMA Nanofibers Under Optimal Conditions

The SP-PMMA nanofibers were prepared by electrospinning technology. The final electrospinning parameters were determined through process optimization. Figure 10 shows the SEM images of the SP-PMMA nanofibers under optimal spinning conditions. It can be seen that the prepared SP-PMMA nanofibers have a smooth surface and a uniform diameter distribution with the average diameter of about 600 nm.

Fig. 9 SEM images of SP-PMMA nanofibers at different propulsion speeds: **a** 0.3 mL/h; **b** 0.4 mL/h; **c** 0.5 mL/h; **d** 0.6 mL/h

Fig. 10 SEM images of SP-PMMA nanofibers at optimal electrospinning conditions (**a**, **b**, **c**, and **d** are photos of the same SP-PMMA nanofibers at different magnification)

3.6 *Fluorescent Spectra of SP-PMMA Nanofibers*

In Fig. 11, curves *a* and *b* are the fluorescence emission spectra of SP-PMMA nanofibers under different excited wavelengths of 367 and 295 nm, respectively. Figure 11a is the spiropyran characteristic emission peak. When excited by ultraviolet light (UV), the closed-ring structure (SP) of spiropyran converts to the open-ring structure (MC). The chromophore in the open-ring structure absorbs excess energy. The fluorescent light is released and shows a bright red glow. Figure 11b shows the

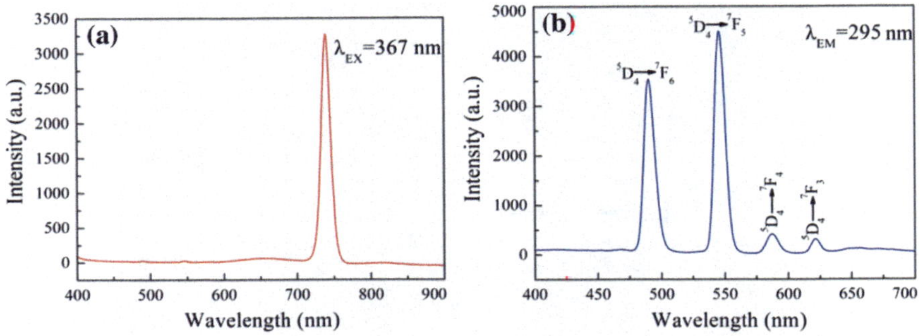

Fig. 11 Emission spectra of SP-PMMA nanofibers under different excited wavelengths

characteristic emission peaks of organic rare earth complexes [$Tb(BA)_3(phen)$, BA = benzoic acid, phen = 1,10-phenanthroline], which correspond to $^5D_4 \rightarrow {}^7F_6$, $^5D_4 \rightarrow {}^7F_5$, $^5D_4 \rightarrow {}^7F_4$, $^5D_4 \rightarrow {}^7F_3$ transition emission at 490, 546, 586, 622 nm, respectively [27], with the strongest emission peak at 546 nm, it shows the bright green light. From Fig. 11, it can be clearly seen that the emission spectra of spiropyran and rare earth complexes [$Tb(BA)_3(phen)$, BA = benzoic acid, phen = 1,10-phenanthroline] under different excitations wavelength do not overlap. This feature provides a prerequisite for multi-base photoluminescent properties of SP-PMMA nanofibers.

3.7 Photochromic Properties of SP-PMMA Nanofibers

The UV absorption is an important method to characterize the photochromic properties of photochromic material [28]. Figure 12 shows the UV absorption spectra of the photochromic and discolored processes of SP-PMMA nanofibers. Figure 12a, b shows the photochromic process of SP-PMMA nanofibers. From Fig. 12a, b, after the SP-PMMA nanofibers are irradiated 2 min by 365 wavelength UV light, the absorption peak intensity of the sample closes to the maximum value. The result indicates that the transformation from closed-loop colorless spiropyran structure to the open-loop colored merocyanine can be completed in 2 min and the SP-PMMA nanofibers have good photochromic function. Figure 12c, d presents the discolored process of SP-PMMA nanofibers. The result indicates that the transformation of MC to SP is slow. After 4 h, the sample still retains obvious purple color. It takes at least 21 h for this sample to convert into completely white color. This result provides the necessary condition for the multi-base photochromic of SP-PMMA nanofibers.

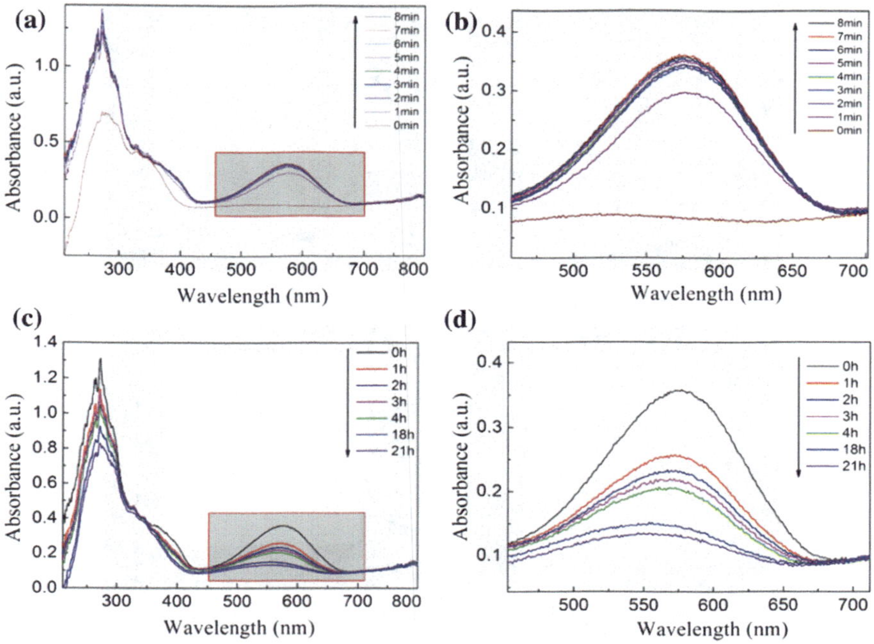

Fig. 12 UV absorption spectra of SP-PMMA nanofibers: (**a**, **b**: the UV absorption spectra of photochromic process of SP-PMMA nanofibers; **c**, **d**: the UV absorption spectra of the colorless process of SP-PMMA nanofibers)

3.8 Multi-base Photochromic and Photoluminescent Properties of SP-PMMA Nanofibers and Nonwoven

The visible product to naked eye of electrospun SP-PMMA nanofibers is white SP-PMMA nonwoven. Figure 13 gives the digital photographs of the SP-PMMA nonwoven under the irradiation of different wavelengths UV lights and far-infrared light. From Fig. 13a, b, the SP-PMMA nanofibers nonwoven show red and green under the excitation of 365 and 254 nm wavelength UV lights, respectively. When the SP-PMMA nonwoven is excited simultaneously by the 365 and 254 nm wavelength UV lights, the sample shows yellow color (Fig. 13c). Figure 13d indicates that the SP-PMMA nanofibers nonwoven show a blue-purple color when the above sample is exposed to visible light. When the sample with blue-purple color is heated by far-infrared light, the SP-PMMA nanofibers nonwoven will return to original white. These results prove that the SP-PMMA nanofibers have the good multi-base photochromic and photoluminescent functions, which have important value for the wide application of nonwoven fabrics.

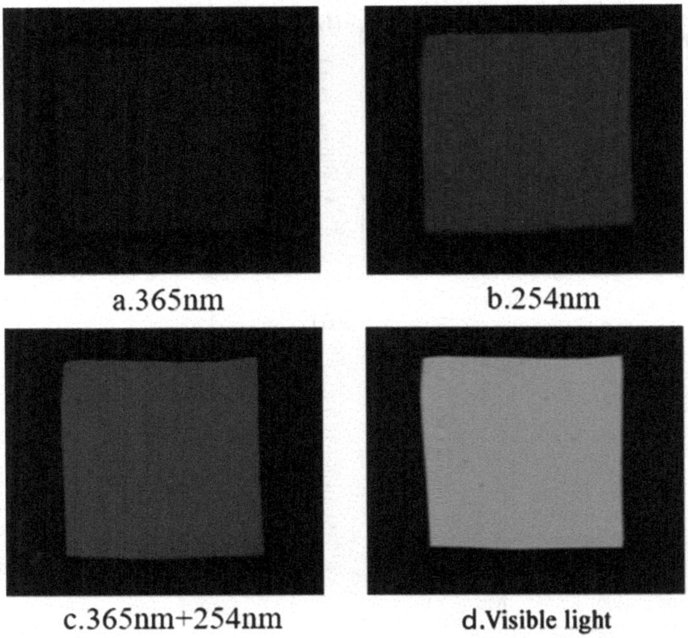

Fig. 13 Photochromic pictures at different ultraviolet wavelengths of nonwoven fabric

4 Conclusions

The polymer with photochromic units was prepared by atom transfer radical polymerization method. The SP-PMMA nanofibers were synthesized by coordination reaction of the polymer with organic rare earth complex and electrospinning technique with optimized spinning concentration of 28%, spinning voltage of 20 kV, receiving distance of 20 cm and the propulsion speed of 0.5 mL/h. The SP-PMMA nanofibers can be converted into red, green, yellow, blue-purple, and white colors under the excitation of 295, 367 nm UV lights and far-infrared light. The SP-PMMA nanofibers exhibit multi-base photochromic and photoluminescent functions, which provide the precondition for wide application of SP-PMMA nanofibers.

Acknowledgements The authors appreciate the financial support provided by the Natural Scientific Foundation of China (51373118) and the Application Fundamental and Advanced Technology Research Proposal Project of Tianjin, China (18JCZDJC38300, 13JCYBJC17200).

Author Information

Notes The authors declare no competing financial interest.

References

1. Y. Zhang, L. Luo, K. Li et al., Large and Reversible in-situ up-conversion photoluminescence modulation based on photochromism via electric-field and thermal stimulus in ferroelectrics. J. Eur. Ceram. Soc. **38**(9), 3154–3161 (2018)
2. V. Ozhogin Ilya, V. Tkachev Valery, S. Lukyanov Boris et al., Synthesis, structure and photochromic properties of novel highly functionalized spiropyrans of 1,3-benzoxazin-4-one series. J. Mol. Struct. **1161**, 18–25 (2018)
3. P. Hao, L. Zhang, J. Shen et al., Structural and photochromic modulation of dimethylbenzotriazolium iodoargentate hybrid materials. Dyes Pigm. **153**, 284–290 (2018)
4. X. Zou, X. Xiao, S. Zhang et al., A photo-switchable and thermal-enhanced fluorescent hydrogel prepared from *N*-isopropylacrylamide with water-soluble spiropyran derivative. J. Biomater. Sci. Polym. Ed. **32**, 1–16 (2018)
5. J. Wei, X. Jiao, T. Wang et al., Electrospun photochromic hybrid membranes for flexible rewritable media. ACS Appl. Mater. Interfaces. **8**(43), 29713–29720 (2016)
6. M. Großmann, A. Klick, C. Lemke et al., Light-triggered control of plasmonic refraction and group delay by photochromic molecular switches. Acs Photonics **2**(9), 1327–1332 (2015)
7. S. Bushuyev Oleksandr, M. Aizawa, A. Shishido et al., Shape-shifting azo dye polymers: towards sunlight-driven molecular devices. Macromol. Rapid Commun. **39**(1), 170–173 (2017)
8. M. Estrader, J. Salinas Uber, A. Barrios Leoni et al., Magneto-optical molecular device: interplay of spin crossover, luminescence, photomagnetism and photochromism. Angew. Chem. Int. Ed. **56**(49), 15622–15627 (2017)
9. K. Kinashi, T. Suzuki, H. Yasunaga et al., Carrier-assisted dyeing of poly(L-lactic acid) fibers with dispersed photochromic spiropyran dyes. Dyes Pigm. **145**, 444–450 (2017)
10. M.N. Liu, X. Yan, M.H. You et al., Reversible photochromic nanofibrous membranes with excellent water/windproof and breathable performance. J. Appl. Polym. Sci. **135**(23), 463–468 (2018)
11. J. Zhang, J. Li, M. Huo et al., Photochromic inorganic/organic thermoplastic elastomers. Macromol. Rapid Commun. **38**(16), 32–36 (2017)
12. K.H. Fries, J.D. Driskell, S. Samanta et al., Spectroscopic analysis of metal ion binding in spiropyran containing polymer thin films. Anal. Chem. **82**(8), 3306–3314 (2010)
13. Y. Fu, C. Fan, G. Liu et al., A colorimetric and fluorescent sensor for Cu^{2+}, and F^-, based on a diarylethene with a 1,8-naphthalimide Schiff base unit. Sensors & Actuators B Chemical **239**, 295–303 (2017)
14. T. Tsuruoka, R. Hayakawa, K. Kobashi et al., Laser patterning of optically reconfigurable transistor channels in a photochromic diarylethene layer. Nano Lett. **16**(12), 7474 (2016)
15. W. Miao, S. Wang, M. Liu, Reversible quadruple switching with optical, chiroptical, helicity, and macropattern in self-assembled spiropyran gels. Adv. Func. Mater. **27**(29), 1701368 (2017)
16. M. Ma, C. Li, D. Shu et al., Synthesis and characterization of bright green terbium coordination complex derived from 1,4-bis(carbonylmethyl)terephthalate: structure and luminescence properties. Spectrochim. Acta Part A Mol. Biomol. Spectrosc. **190**, 68–75 (2018)
17. B. Gao, F. Li, R. Zhang et al., Preparation of aromatic carboxylic acid-functionalized polysulfone and preliminary exploration of florescence emission character of formed polymer-rare earth complexes. Synth. Met. **162**(5–6), 503–510 (2012)
18. C. Ventura, R. Byrne, F. Audouin et al., Atom transfer radical polymerization synthesis and photoresponsive solution behavior of spiropyran end-functionalized polymers as simplistic molecular probes. J. Polym. Sci., Part A: Polym. Chem. **49**(16), 3455–3463 (2011)
19. H. Liu, J. Hu, X. Yang et al., Preparation and characterization of dual-responsive spiropyran-based random polymer brushes via surface-initiated atom transfer radical polymerization. Des. Monomers Polym. **19**(2), 193–204 (2016)
20. C. Pauly Anja, Schöller Katrin, Baumann Lukas et al., ATRP-based synthesis and characterization of light-responsive coatings for transdermal delivery systems. Sci. Technol. Adv. Mater. **16**(3), 272–275 (2015)

21. D. Shu, P. Xi, S. Li et al., Morphologies and properties of PET nano porous luminescence fiber: oil absorption and fluorescence-indicating functions. ACS Appl. Mater. Interfaces. **10**(3), 2828–2836 (2018)
22. F. Khakzad, A.R. Mahdavian, H. Salehi-Mobarakeh et al., Redispersible PMMA latex nanoparticles containing spiropyran with photo-, pH- and CO_2-responsivity. Polymer **101**, 274–283 (2016)
23. Y. Wang, C.Y. Hong, C.Y. Pan, Spiropyran-based hyperbranched star polymer: synthesis, phototropy, FRET, and bioapplication. Biomacromol **13**(8), 2585–2593 (2012)
24. L. Li, Z. Jiang, M. Li et al., Hierarchically structured PMMA fibers fabricated by electrospinning. RSC Advances **4**(95), 52973–52985 (2014)
25. F.L. Zhao, L. Gao, C.H. Wang et al., Electrospinning preparation and properties of Tb(phen)L3/PMMA rare earth luminescent fibers. Appl. Mech. Mater. **618**, 94–99 (2014)
26. M. Obaid, A.M. Barakat Nasser, O.A. Fadali et al., Effective and reusable oil/water separation membranes based on modified polysulfone electrospun nanofiber mats. Chem. Eng. J. **259**(1), 449–456 (2015)
27. F. Zhao, P. Xi, H. Xia et al., Luminescent polymethacrylate composite nanofibers containing a benzoic acid rare earth complex: morphology and luminescence properties. J. Alloy. Compd. **641**, 132–138 (2015)
28. M. Breslin Vanessa, Miguel A. Garcia-Garibay, Transmission spectroscopy and kinetics in crystalline solids using aqueous nanocrystalline suspensions: the spiropyran-merocyanine photochromic system. Cryst. Growth Des. **17**(2), 637–642 (2017)

Network Structures and Thermal Characteristics of Bi_2O_3–SiO_2–B_2O_3 Glass Powder by Sol-Gel

Peixian Li, Gecheng Yuan, Zhenghua Lu, Qian Li and Qiguang Wu

Abstract Glass powder of Bi_2O_3–SiO_2–B_2O_3 was prepared by Sol-Gel method, and the powder was heated to the temperature range from 200 to 800 °C to study the network structures formed during heat treatment. The effect of structures change of the glass powders on the transition temperature and sintering softening properties was analyzed. The results indicate that Bi^{3+} get into the network structure with the rising of heat treatment temperature, [BiO_6] octahedral and [BiO_3] triangle, [BO_4] tetrahedron and [BO_3] triangle connected with [SiO_4] tetrahedron separately in the way of vertex connecting to build the network structures. The O1s and Bi4f binding energies increase gradually while the B1s binding energies decrease, which strengthen the stability of glass structure. This causes an increase in the transition temperature and a decrease in the wettability of the glass powder. The glass powders treated at 600 °C have excellent sintering properties. The glass transition temperature (T_g) is about 542, and the thermal expansion coefficient (25–300 °C) is close to 6.57×10^{-6}/°C.

1 Introduction

Bi_2O_3–SiO_2–B_2O_3 glass powder is a research hot-spot of lead-free glass system. It has broad application prospects in electronic packaging, electronic paste, and so on

---

The original version of this chapter was revised: correct acknowledgement has now been updated. The correction to the book is available at https://doi.org/10.1007/978-981-13-5947-7_25

P. Li · G. Yuan (✉) · Z. Lu · Q. Li
School of Materials and Energy, Guangdong University of Technology, Guangdong, Guangzhou 510006, China
e-mail: gchyuan@gdut.edu.cn

P. Li
e-mail: lipeixianx@163.com

Q. Wu
Analysis and Test Center, Guangdong University of Technology, Guangzhou 510006, China

Y. Han (ed.), *Physics and Techniques of Ceramic and Polymeric Materials*, Springer Proceedings in Physics 216,
https://doi.org/10.1007/978-981-13-5947-7_24

[1]. Bi_2O_3 is an important component of bismuthate glasses. The structural properties such as coordination state and bond length of bismuth ions are the key factors affecting the structure and material properties of boron oxide and silicon oxide networks [2, 3]. Some literatures have shown that the spatial network structure of bismuthate glasses consists of Bi^{3+} with large radius and low field strength combined with a small amount of traditional glass network-forming bodies (such as B^{3+} and Si^{4+}) [4]. However, some scholars believe that Bi^{3+} will undergo polarization deformation and form a spatial network structure in the form of asymmetric [BiO_6] octahedron [5]. B_2O_3 is a glass network-forming body, and its high glass forming ability is derived from a complex stacked structure, and the amorphous random network structure is formed by a B-O ring and a BO_3 triangular chain through a B-O-B link. Therefore, when Bi_2O_3 and B_2O_3 together form a glass, Bi^{3+} with high ion polarization and multiple coordination numbers is likely to form a more complex amorphous structure with a small amount of B^{3+}, and the resulting network structure will directly affect the thermal performance of the glass powder. At present, the preparation of glass powder is mostly carried out by melt quenching method, and the structure research of the glass powder prepared by the method has been reported, while the structural analysis of the glass powder prepared by the new preparation method such as the sol-gel method is rarely systematically studied, especially in the process of heat treatment during the preparation process, the structural changes of the glass powder need to be studied in depth [6–8]. Therefore, it is necessary to have an in-depth understanding of the relationship between the structural change of the glass powder and its thermal properties. In this paper, prepared by sol-gel method, a Bi_2O_3–SiO_2–B_2O_3-based glass powder, the interconnection and network formation mechanism of various network groups in glass at different heat treatment temperatures was studied. Based on this, a glass network structure model was constructed and the influence of network structure changes on glass powder transformation temperature (T_g) and sintering softening characteristics was analyzed.

2 Experiment

2.1 Sample Preparation

$Bi(NO_3)_3 \cdot 5H_2O$, TEOS, H_3BO_3, $Zn(NO_3)_2 \cdot 6H_2O$, $Al(NO_3)_3 \cdot 9H_2O$ are used as precursors; deionized water, ethanol and nitric acid are used as solvents and catalysts, respectively. According to the glass composition oxide ratio (quality percentage) 60%Bi_2O_3, 10%SiO_2, 20%B_2O_3, 7%ZnO and 3%Al_2O_3, the mass of the raw material is converted into a solution, mixed in a certain order and mechanically stirred, and the mixed solution is aged in a constant temperature water bath at 60 °C for a certain period of time to obtain a transparent wet gel. The wet gel was dried at 100 °C for 1 h in a dry box, and an appropriate amount of dry gel was selected and heat treated in a muffle furnace at 200, 400, 600 and 800 °C for 1 h and then subjected to mechanical ball milling for 20 h to obtain a glass powder sample. Several powder samples were taken out for characterization. The remaining powders were made into a cylindrical sample for sintering test and thermal expansion performance test using a pressure tester and a specific mold. The cylindrical sample has a diameter (Φ) of 8 mm and the height (h) of 12 mm.

2.2 *Sample Characterization*

The morphology of the glass powder was analyzed by Hitachi S-4300N SEM and JEM-1400 TEM analyzer. The glass network structure was analyzed by ESCALAB-250 XPS, Nicolet 380 FT-IR and inVia Raman; Y- Model 2000 XRD was used to analyze the crystallization characteristics of glass powder samples at different heat treatment temperatures. The glass transition characteristics and thermal expansion characteristics of glass powder samples in various temperature ranges were tested by 3SDT-2960 DSC-TG and NETZSCH DIL402PC thermal dilatometer. The button sintering experiment is used to test the wettability of the glass powder and the matrix material, and the contact angle θ is measured according to the principle of the sessile drop method with (1).

$$\theta = 2tg^{-1}\left(2h \cdot d^{-1}\right) \quad (1)$$

where, h, d are the height and diameter of the glass after sintering, respectively.

3 Results and Discussion

3.1 *Network Structure Analysis*

Figure 1a shows the IR absorption spectrum of the powder after heat treatment at different temperatures. With the treatment temperature increases, the OH^{-1} or H_2O vibration in the powder decreases at 3450 and 1600 cm^{-1} [9]. The intensity of B–O bond stretching vibration peak in [BO_4] at 1182 cm^{-1} decreases, while the vibration intensity of B–O–Si structure increases at 1370 cm^{-1}. It is indicated that as the heat treatment temperature increases, the [BO_4] tetrahedron is connected with the [SiO_4] tetrahedron apex angle to share the O atom to form the Si–O–B structure [10]. The absorption intensity caused by the vibration of the [BiO_3] triangular structure at 900 and 860 cm^{-1} gradually decreases, while the vibration intensity of the Si–O–Bi structure at 670 cm^{-1} gradually increases. It shows that Bi^{3+} is gradually added to the Si–O–Si network and participates in the construction of the glass network to form the Si–O–Bi structure [11]. As [BO_4] is connected to [SiO_4] and Bi^{3+} ions are continuously added to the Si–O–Si network, the symmetry of the [SiO_4] tetrahedron

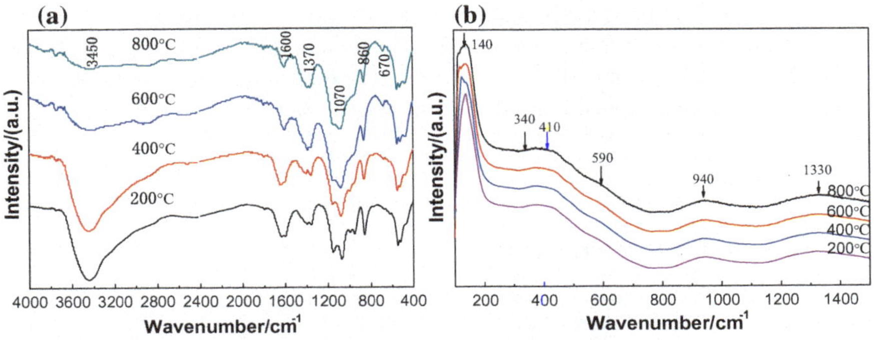

Fig. 1 IR (**a**) and Raman (**b**) spectrum of powder treated at different temperature

is destroyed, that results in absorption intensity structure of Si–O–Si asymmetric stretching vibration at 1070 cm^{-1} gradually increased.

Figure 1b shows the Raman spectrum of the powder after heat treatment at different temperatures. The most intense peaks occurred at 140 cm^{-1}, it is the characteristic vibration peak of [BiO_6] octahedron and [BiO_3] triangle caused by Bi^{3+} ion as a network forming body entering the glass structure [11]. The scattering peaks at 305–325 and 410–420 cm^{-1} are Bi–O–Bi symmetric stretching vibration modes in [BiO_6] and [BiO_3], respectively [12]. As the heat treatment temperature increases, the scattering intensity of the glass powder increases at the two locations. It shows that the number of [BiO_6] and [BiO_3] increases with Bi^{3+} ions entering the glass network. The vibration peak near 575–590 cm^{-1} is mainly due to the stretching vibration of $Bi–O^-$ non-bridge oxygen bond in [BiO_6], and temperature has little effect on $Bi–O^-$ non-bridged oxygen bonds in glass [13]. The scattering peak at 1250–1330 cm^{-1} is the stretching vibration of B–O–B in the [BO_3], and the intensity of the scattering peak gradually decreases with the increase of temperature. The reason is that the temperature rise causes the [BO_3] triangle to break into a chain structure, and the [BO_3] triangle is connected with the [SiO_4] tetrahedron apex angle to share the O atom to form the Si–O–B structure, and the vibration directed low wave direction, it is indicated that the [BO_3] group may be attached to the [BiO_6] or [BiO_3] group with a lower wave number due to the high charge of Bi^{3+} and the strong polarization. When the symmetric stretching vibration peak of Si-O-Si in the [SiO_4] is 940 cm^{-1}, the scattering peak intensity decreases with increasing temperature [11]. It shows that Bi^{3+} enters the Si–O–Si network structure and participates in the construction of the glass network. At the same time, [BO_3] is connected with [SiO_4], which destroys the symmetry of the [SiO_4] tetrahedron, which is consistent with the IR spectrum analysis.

Figure 2a is an O1s XPS of the powder after heat treatment at 200 °C. The peak of O1s has a width of about 6 eV, which can be divided into bridge oxygen (BO) and non-bridge oxygen (NBO), and the non-bridge oxygen peak area is smaller than the bridge oxygen peak area [14]. It is indicated that the oxygen in the glass structure is mainly in the form of Si–O–Si. At the same time, part of Bi^{3+} in the glass

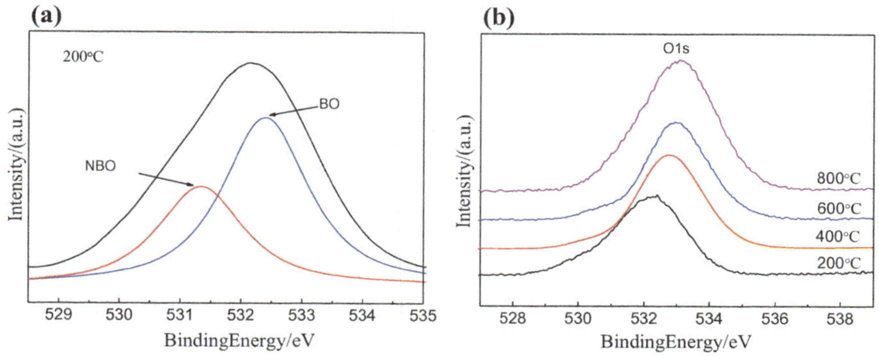

Fig. 2 O1s XPS of powder treated at 200 °C (**a**) and different temperature (**b**)

enters the glass network structure after heat treatment, and the [BiO_6] or [BiO_3] is connected with the [SiO_4] to form Si–O–Bi. Figure 2b shows the O1s XPS of the powder after heat treatment at different temperatures. With the increase of heat treatment temperature, the electron binding energy of O1s increased from 532.3 to 533.1 eV, and the oxygen ion concentration also increased gradually, indicating that the chemical bonds formed by other atoms connected with O are large, the cations are enhanced by oxygen ion binding, and glass network structure is more compact [11, 14].

Figure 3a shows the XPS of Bi4f of powder after heat treatment at different temperatures. As the temperature increases, the binding energy of $Bi4f_{7/2}$ increases from 159.5 to 159.9 eV, and the binding energy of $Bi4f_{5/2}$ increases from 164.8 to 165.3 eV. This may be due to the increase of oxygen ion concentration in the glass. The $6s^2$ lone pair of electrons of some Bi^{3+} ions are strongly electrostatically repelled by oxygen ions, which makes the Bi^{3+} ions become Bi^{5+} ions, which reduces the electron density of Bi–O bonds [14]. In addition, since the Si–O bond length in the glass network structure is 1.61 Å, the temperature rises, and Bi^{3+} gradually enters the glass network structure, and the Bi–O bond length of the Si–O–Bi structure is between 2.2 and 2.8 Å. Therefore, the formed Si–O–Bi structure is increased in electrostatic repulsion due to the increased cation distance between silicon and germanium, and the stability of the glass network structure is enhanced.

Figure 3b shows the XPS of B1s of the powder after heat treatment at different temperatures. With increasing temperature, the binding energy of B1s decreases from 192.5 to 192.4 eV. This is probably because of the [BO_3], [BO_4] and [SiO_4] form Si–O–B structures. Because the electronegativity of B is larger than that of O and Si, the electrons tend to shift to the large electronegative atoms which lead to the increase in the density of the electron cloud around B, the increase of the shielding effect and the decrease of the corresponding electron binding. Moreover, since the B–O bond energy is 119 kcal/mol, which is larger than the Si–O bond energy of 106 kcal/mol, the formed Si–O–B structure is more stable.

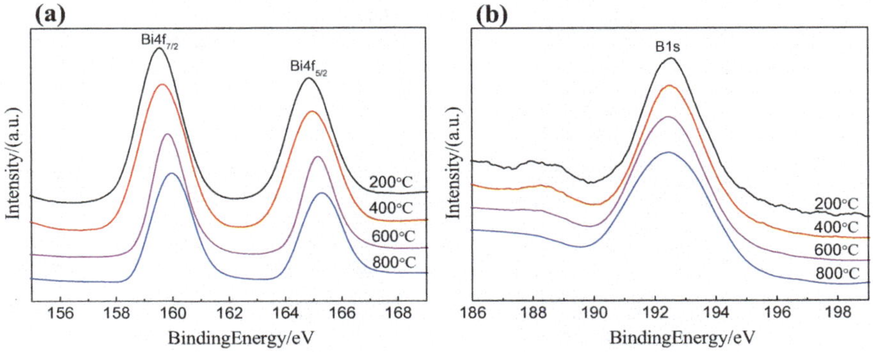

Fig. 3 Bi4f (**a**) and B1s (**b**) XPS of powder treated at different temperature

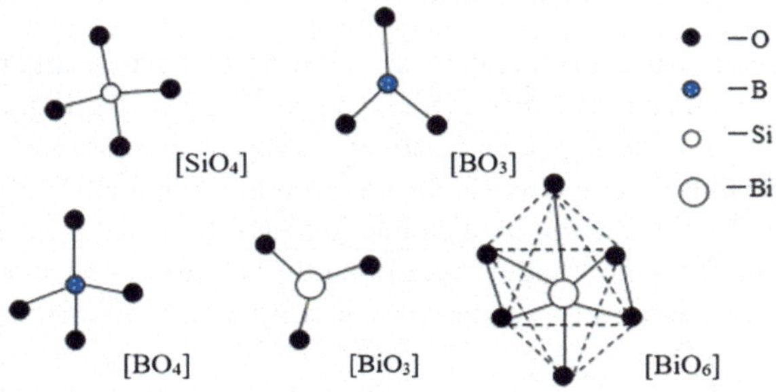

Fig. 4 Structural units of glass powder

The above IR, Raman spectrum and XPS analysis show that the units forming glass network structure are mainly like [BiO_6], [BiO_3], [BO_4], [BO_3] and [SiO_4] in Fig. 4. As the main component of glass, Bi_2O_3 exists first in the form of modifier in the glass network. The network structure of the glass is mainly constructed by the [SiO_4] tetrahedron in SiO_2 and the [BO_4] tetrahedron in B_2O_3 and the [BO_3] in the way of the top angle, as shown in Fig. 8.

After heat treatment at different temperatures, Bi^{3+} gradually enters the glass network, destroying the original glass network structure. The [SiO_4], [BO_4] and [BO_3] body, respectively, form a glass network structure with the formed [BiO_6] and [BiO_3] reconnected at the vertex angle. As shown in Fig. 5a, the [SiO_4], the [BO_4] and the [BO_3] are constructed with the [BiO_6] and the [BiO_3], which are connected to the top angle, respectively, as shown in Fig. 5b.

3.2 *Thermal Performance*

Figure 6a shows the XRD of the powders after heat treatment at different temperatures. After heat treatment at different temperatures, the diffraction peaks of the

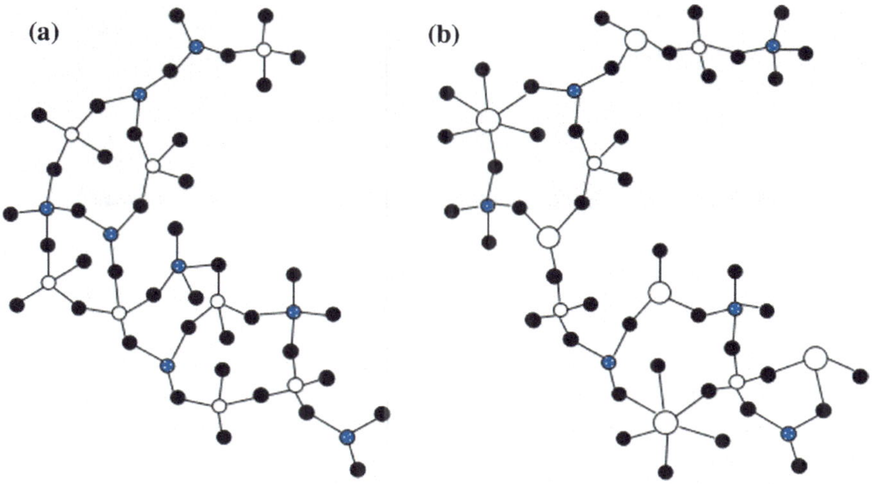

Fig. 5 Structural model of original glass (**a**) and glass powder after heat treatment (**b**)

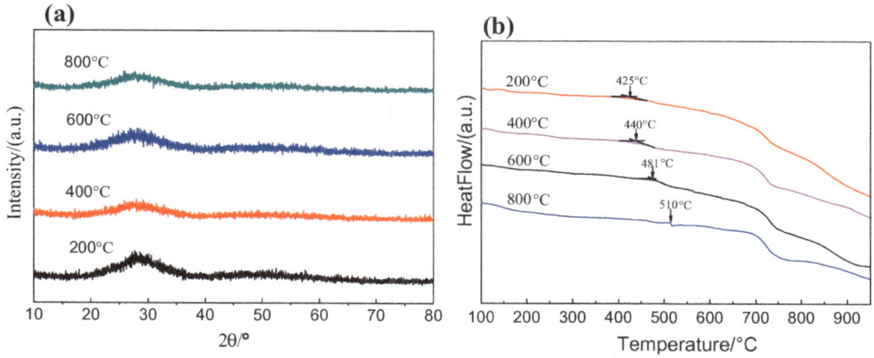

Fig. 6 XRD spectrum (**a**) and DSC curves (**b**) of powder treated at different temperature

patterns did not change significantly. It has always been characterized by the widening and dispersion characteristics of typical amorphous materials [15]. That is, the glass state powders of the system can be successfully prepared by sol-gel method.

The transition temperature (T_g) is an important characteristic temperature of glass [5]. The size of T_g is closely related to the bond strength between atoms and the network structure of glass. Figure 6b is the DSC curve of the glass powder after heat treatment at different temperatures. With the increase in heat treatment temperature, the T_g of glass powder increased from 425 to 510 °C. It shows that the network structure of glass increases gradually and the aggregation degree increases, which is corresponding to the above network structure analysis. The melting endothermic peaks of the powders after heat treated at different temperatures were between 700

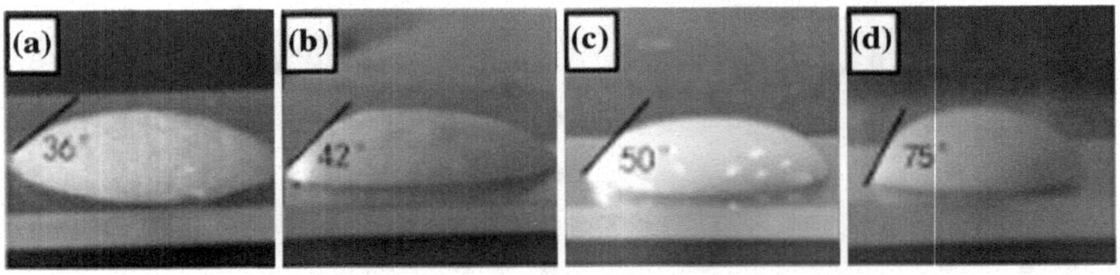

Fig. 7 Photographs showing the softening state of sintered cylindrical samples of powder treated at different temperature. **a** 200 °C; **b** 400 °C; **c** 600 °C; **d** 800 °C

and 750 °C. It is indicated that when the glass powder is sintered in this temperature range, the viscosity of the glass is insufficient to support the glass to cause a large shape change. In addition, there is no crystal exothermic peak in the whole test temperature range.

Photographs showing the softening state of sintered glass samples at 725°C with heat-treated glass powders under different temperatures is Fig. 7. The cylindrical pattern of glass powder becomes hemispherical or even dispersed. Combined with the above DSC analysis, it is known that the inner surface of the glass powder cylinder is close to each other and bonded, increased fluidity, resulting in surface profile deformation [16]. The glass powder after the heat treatment may have better wetting with the substrate at different temperatures, sintered glass frit powder preferably wettability of 200 °C, but the surface is rough and porous after sintering, resulting in poor density. As the heat treatment temperature increases, the contact angle increases from 36° to 75°. The wettability has a certain degree of decline, but the density is increased after sintering, and the surface has a clear glass luster. Therefore, considering the comprehensive consideration, the glass powder after heat treatment at 600 °C has good wettability, density and surface gloss.

Figure 8a is the SEM morphology of the glass powder after heat treatment at 600 °C for 20 h after ball milling. The particle size of powder is less than 1 nm and mainly spherical like. Further, the glass powder is observed by TEM. As shown in Fig. 8b, the surface of the glass powder is covered with nanoparticles of about 10–20 nm [15]. This is the embodiment of colloidal particles in powder. During the gel formation process, the colloidal particles are continuously increased and grown to form a gel. After drying, the nanoparticles are stored in the glass yet. In the ball milling process, the nanoparticles are adsorbed on the surface of the glass powder and piled up layer by layer. Therefore, the specific surface area of the prepared glass powder is large, and the surface energy is higher. In the sintering process, it is beneficial to improve the fluidity and the densification of the sample. This is also reflected in the above sintering experiments.

Figure 9 is a graph showing the thermal expansion coefficient of the glass powder pattern after heat treatment at 600 °C. The curve shows an upward sharp turn near 485 °C, corresponding to the T_g, a peak occurs at about 542 °C, corresponding to the dilatometric softening point temperature of glass (T_d). The coefficient of thermal

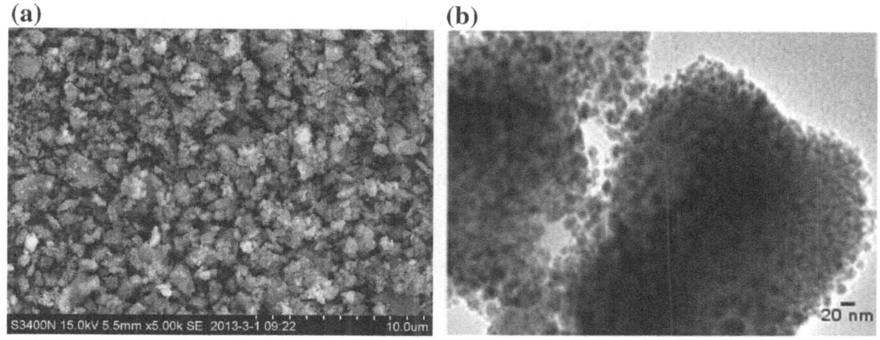

Fig. 8 SEM (**a**) and TEM (**b**) image of powder treated at 600 °C

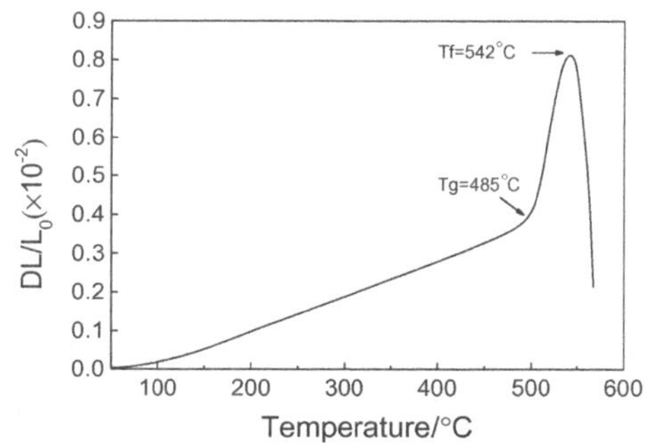

Fig. 9 Thermal expansion curve of powder treated at 600 °C

expansion (25–300 °C) α was calculated to be about 6.57×10^{-6}/°C. Therefore, the glass powder after heat treatment at 600 °C has a stable network structure and a low softening temperature, does not undergo crystallization during sintering, and has good wettability with the substrate, while can be applied to the sealing temperature of the base material at about 725 °C. The coefficient of thermal expansion (25–300 °C) is about 6.57×10^{-6}/°C for fluxing and sintering.

4 Conclusions

(1) Within the test temperature range, as the heat treatment temperature increases, Bi^{3+} gradually enters the glass network, [BiO_6] and [BiO_3], [BO_4] and [BO_3] connected with [SiO_4] separately in the way of vertex connecting to build the network structures.

(2) The structures of Si–O–Bi and Si–O–B of glass powder form gradually during heat treatment. With the increase of heat treatment temperature, the electron binding energy of O1s and Bi4f increases and the electron binding energy of

B1s decreases gradually. Therefore, the stability of the glass network structure is enhanced.

(3) The change of the glass powder network structure is the main reason that the T_g increases and the wettability decreases with the heat treatment temperature. The powder after heat treatment at 600 °C has good sintering performance and softening temperature. The coefficient of thermal expansion (25–300 °C) α is about 6.57×10^{-6}/°C.

Acknowledgements This work was financially supported by Guangdong Provincial Natural Science Fund of China (Nos. 2006B14701003) and Guangzhou Science and Technology Project of China (Nos. 201510010034).

References

1. Y.U. Xiaojun, Z.H.U. Lihui et al., Effects of Bi_2O_3 on structure and properties of Al_2O_3-ZnO-Bi_2O_3-B_2O_3 low-melting glasses. Electron. Compon. Mat. **32**(9), 12–14 (2013)
2. H.E. Feng, D.E.N.G. Dawei, W.A.N.G. Jun, Effect of Bi_2O_3 contents on sintering property of Bi_2O_3-ZnO-B_2O_3 system low-melting electronic sealing glass. J. Wuhan Univ. Technol. **31**(22), 1–4 (2009)
3. Z. Hongping, Z. Renjie, Effects of CuO on structure and heat treatment of Bi_2O_3-B_2O_3-ZnO glasses. J. Ceram. **31**(4), 569–574 (2010)
4. J. Cheng, F. Chen, S. Dai, et al., Vitreous network formation and optical characteristics of glasses within Bi_2O_3—B_2O_3 binary system. J. Chin. Ceram. Soc. **41**(4), 475–479 (2013)
5. Y. Huang, Y. Li, J. Wang, et al., Network structures and characteristics of Bi_2O_3-ZnO-B_2O_3 Ternary system glasses. J. Chin. Ceram. Soc. **43**(7), 998–1001 (2015)
6. S. Shruti, A.J. Salinas, G. Malavasi et al., Structural and in vitro study of cerium, gallium and zinc containing Sol-Gel bioactive glasses. J. Mater. Chem. **22**(27), 13698–13706 (2012)
7. J. Chen, W. Que, Y. Xing et al., Molecular level-based bioactive glass-poly (caprolactone) hybrids monoliths with porous structure for bone tissue repair. Ceram. Int. **41**(2), 3330–3334 (2015)
8. B.B. Das, A. Srinivassan, M. Yogapriya, et al., Sol–gel synthesis and characterization of xCuO-(1-x)Bi_2O_3 ($0.15 \leq x \leq 0.55$) glasses by magnetic and spectral studies. J. Non-Cryst. Solids **427**, 146–151 (2015)
9. M. Wang., Investigation of the structure evolution process in sol–gel derived CaO-B_2O_3-SiO_2 glass ceramics. J. Non-Cryst. Solids 357, 1160–1163 (2011)
10. X. Zhu, C. Mai, M. Li, Effects of B_2O_3 content variation on the Bi ions in Bi_2O_3- B_2O_3- SiO_2 glass structure. J. Non-Cryst. Solids **388**, 55–61 (2014)
11. H. Fan, Infrared, Raman and XPS spectroscopic studies of Bi_2O_3-B_2O_3-GeO_2 glasses. Solid State Sci. **12**(4), 541–545 (2010)
12. Y. Zhang, Y. Yang, J. Zheng et al., Effects of oxidizing additives on optical properties of Bi_2O_3-B_2O_3-SiO_2 glasses. J. Am. Ceram. Soc. **91**(10), 3410–3412 (2008)
13. I. Ardelean, S. Cora, FTIR and Raman investigations of MnO-B_2O_3-Bi_2O_3. Optoelectron. Adv. Mater. **12**(2), 239–243 (2010)
14. H.W. Nesbitt, G.M. Bancroft, et al., Bridging, non-brifging free (O_2-) oxygen in Na2O-SiO2 glasses. J. Non-Cryst. Solids **1**(357), 173–175 (2011)
15. L.E. Yingfeng, Y.U.A.N. Gecheng, L.I. Qian et al., Thermal properties of Bi_2O_3-SiO_2-B_2O_3-Zn O-Al_2O_3 glass powders prepared by sol-gel method. China Powder Sci. Technol. **23**(1), 85–87 (2017)

16. D.I.N.G. Kanjunjie, C.H.E.N. Nan, Preparation, structures, thermal properties and sintering behaviors of B_2O_3-SiO_2-ZnO-BaO-Al_2O_3 glass. J. Wuhan Univ. Technol. Mater **31**(6), 1323–1328 (2016)

Correction to: Network Structures and Thermal Characteristics of Bi_2O_3–SiO_2–B_2O_3 Glass Powder by Sol-Gel

Peixian Li, Gecheng Yuan, Zhenghua Lu, Qian Li and Qiguang Wu

Correction to:
Chapter "Network Structures and Thermal Characteristics of Bi_2O_3–SiO_2–B_2O_3 Glass Powder by Sol-Gel" in:
Y. Han (ed.), *Physics and Techniques of Ceramic and Polymeric Materials*,
Springer Proceedings in Physics 216,
https://doi.org/10.1007/978-981-13-5947-7_24

The original version of this chapter was inadvertently published with incorrect grant number in the acknowledgement section. The grant number of Guangzhou Science and Technology Project of China has been changed from 2015110010034 to 201510010034.

The updated version of this chapter can be found at
https://doi.org/10.1007/978-981-13-5947-7_24

Y. Han (ed.), *Physics and Techniques of Ceramic and Polymeric Materials*,
Springer Proceedings in Physics 216,
https://doi.org/10.1007/978-981-13-5947-7_25

Index

Y. Han (ed.), *Physics and Techniques of Ceramic and Polymeric Materials*, Springer Proceedings in Physics 216, https://doi.org/10.1007/978-981-13-5947-7